BETWEEN PARADISE and PERIL

The Natural Disaster History of the Monterey Bay Region

Gary Griggs

Monterey Bay Press

Published by Monterey Bay Press (www. montereybaypress.com), Santa Cruz County, California.

Photographs
Top Cover: Flood damage along the bank of Aptos Creek on Moosehead Drive in January 1982 by Gary Griggs © 1982.
Bottom Cover: Landslide and flood damage along Corralitos Creek in Eureka Canyon in 1986 by Gary Griggs © 1986.
Back Cover: Author photo by Deepika Shrestha Ross © 2015.
All other photographs © Gary Griggs unless noted otherwise.

Colophon
Cover and Book Design: D. Shrestha Ross Studio, using Adobe® InDesign® and typefaces Adobe Caslon Pro for the body text, and Franklin Gothic Medium for the titles and headings.

For my five children, *Joel, Amy, Shannon, Callie and Cody,*
who all grew up in Santa Cruz County, with the hope that they will never
suffer the consequences of any geologic hazards, and that as a state, nation,
and planet that we can soon begin to bring climate change under control.

"Living in the Monterey Bay region means living in a constant state of contradiction. We live in one of the most bountiful, beautiful places on Earth, and yet Mother Nature brings ever-present challenges of flood, fire, drought, landslides, tsunami, and seismic activity. ***Between Paradise and Peril*** *captures this balance in a complete, understandable and entertaining way."*

– John Laird
California Secretary for Natural Resources

"Gary Griggs has provided a timely and beautifully crafted work. While it focuses on a history of disasters and the need to be ever vigilant, ***Between Paradise and Peril*** *offers a view of the natural, ecological, and human history of our region that serves to remind us of the paradise we are so fortunate to inhabit and must work to protect."*

– Senator Bill Monning
California State Senate Majority Leader, 17th Senate District

"Surrounded daily by the stunning natural beauty of Monterey Bay, it is all too easy to forget about the dark side of Mother Nature. ***Between Paradise and Peril*** *is a compelling and important reminder of the real danger we put ourselves in by living in our little slice of paradise. Dr. Gary Griggs has written a fascinating book that needs to be read by all of us who live in or visit the Monterey Bay. After all, we need to remember that paradise can turn on us at any moment."*

– Neal Coonerty
former Mayor of Santa Cruz & former Santa Cruz County Supervisor

Contents

Other Books by Gary Griggs:

The Earth and Land Use Planning
Duxbury Press, 1977
with John Gilchrist

Geologic Hazards, Resources and Environmental Planning
Wadsworth Publishing Company, 1983
with John Gilchrist

Living with the California Coast
Duke University Press, 1985
Author and Editor with Lauret Savoy

Living with the Changing California Coast
UC Press, 2005
Author and Editor with Lauret Savoy and Kiki Patsch

Santa Cruz Coast: Then and Now
Arcadia Publishing, 2006
with Deepika Shrestha Ross

Introduction to California's Beaches and Coast
UC Press, 2010

California Coast from the Air: Images of a Changing Landscape
Mountain Press, 2014
with Deepika Shrestha Ross
and photography by Ken and Gabrielle Adelman

Coasts in Crisis: A Global Challenge
UC Press, 2017

The Edge: The Pressured Past and Precarious Future of California's Coast
Craven Street Books, 2017
with Kim Steinhardt

Acknowledgements

While almost every book has only a single name on the cover, there are usually many more people involved in making that book a reality. I want to acknowledge the many conversations, email exchanges, presentations, field trips and hikes over the past 35 years with my good friend and colleague, Sandy Lydon ("the History Dude"), which added significantly to my own historic knowledge of the Monterey Bay Region, and allowed me to fill in important pieces of this story. Thank you Sandy for all of our many collaborations, conversations and exchanges over these years. And of course, any errors in history are mine alone.

To the many photographers and graphic designers, as well as the various collections and organizations (UCSC Special Collections, UCSC Map Library, the Capitola Natural History Museum, the Santa Cruz Museum of Natural History, the McPherson Art and History Museum, the Santa Cruz Public Library System, the *Santa Cruz Sentinel*, the Santa Cruz Beach Boardwalk Archives, the California Coastal Records Project (Ken and Gabrielle Adelman), the Library of Congress, the Carnegie Commission Report on 1906 Earthquake, the United States Geological Survey (USGS), UC Press, UC Berkeley Seismological Laboratory, UC Berkeley Bancroft Library, the Advocates for the Forest of Nisene Marks, the Dudley Knox Library at the Naval Postgraduate School, the Meriam Library at California State University Chico, the United States Navy, *National Geographic*, the National Oceanic and Atmospheric Administration (NOAA), the National Park Service, UC Water Resources Center Archives, the Orville Magoon private collection, the Santa Cruz Port District, the United States Army Corps of Engineers, the Santa Cruz Economic Development Department, California American Water (CalAm), the *New York Times*, the *San Jose Mercury*, the Santa Cruz City Water Department, the United States Global Change Research Program, the Monterey Bay Aquarium Research Institute, and AVISO CNES Data Center) who generously allowed me to use their photographs and illustrations, I am most grateful.

I also owe a debt of gratitude to the newspapers and reporters of the Monterey Bay region who wrote and published the original stories that allowed me to dig into the history of the area's natural disasters. In addition, my thanks to Heather Moffatt, former Director of the Santa Cruz Natural History Museum for her initial editing of the manuscript.

Last but definitely not least, I thank my wife and partner, Deepika Shrestha Ross, for her generosity, patience, and considerable design skills which brought together my writing and the many photographs and illustrations together in this book.

Foreword

Bruce McPherson
Santa Cruz County Supervisor, 5th District

Living in the Monterey Bay Region, we are blessed with abundant natural beauty and endless opportunities to enjoy the wonders around us. In the span of just a few hours, you can ride a wave in one of our world-class surf breaks, hike through dense redwood forests or sweeping coastal prairies, and watch a monarch butterfly flutter toward a clear blue sky.

While we are fortunate to live in paradise, we mustn't forget that it is a fragile paradise – one prone to disaster in the face of stormy winters, tinder-dry summers and, irrespective of those extreme conditions, ever-present fault lines and coastal erosion that shifts the Earth beneath us. As we appreciate the riches of nature, so must we acknowledge and prepare for a variety of devastating events that often carry no immediate warning, although the threat is always there.

Our homes, waterways, infrastructure, and sadly sometimes our lives, hang in the balance as we try our best to predict and prevent catastrophe through our deep commitment to slowing climate change. As do many other lifelong residents of our region, my memories of flash floods, searing wildfires, forceful mudslides and one massive earthquake can be triggered in an instant when I see photographs that illustrate our region's vulnerabilities.

I remember as an 11-year-old boy during the 1955 Christmas Floods stacking sandbags to protect the *Santa Cruz Sentinel*, which my family owned at the time. In 1982, while serving as editor of the *Sentinel*, I marveled once more at the community's heroism and compassion during the Love Creek mudslide disaster. And it was just seven years later that the Loma Prieta earthquake ravaged downtown Santa Cruz, Watsonville and communities in the Santa Cruz Mountains. I was watching the World Series in San Francisco when Candlestick Park shook, and many hours passed before I could get home to take a first-hand look at the destruction. And when I think about the most urgent danger we face now what comes to mind instantly is how the Martin, Summit and Bear Creek fires roared through our bucolic mountain reaches.

As expertly and engagingly explored in this book, ***Between Paradise and Peril: The Natural Disaster History of the Monterey Bay Region***, our history of environmental peril is not only long and storied, but also doomed to repeat itself. Over a career spanning five decades, author Gary Griggs, a UCSC Distinguished Professor of Earth and Planetary Sciences, has studiously documented our changing landscapes and marine sanctuary, challenging us to assume the responsibility of understanding our surroundings and the collective accountability required to better steward them.

It's impossible to know exactly what our region will look like 100 years or even 1,000 years from now. But due to the work of Gary and many other dedicated scientists, we have a clearer picture than did past generations that our small corner of the world – as we know it and see it every day – does not come with a lifetime guarantee. The important history lessons contained in this book are as inspiring as they are instructive about our resilience as a community and our determination to adapt. Gary is to be commended for making this clarion call not only to learn from our past but to turn knowledge into action.

Preface

Some people have accused me of loving disasters. While that's not completely true, I will admit that during my first half century living in the Monterey Bay area, and experiencing an almost unending series of natural disasters – whether floods or droughts, earthquakes or landslides, or coastal storms and shoreline erosion – that these recurring events have been pretty exciting for an Earth scientist to witness and study.

While it might seem that we have had more than our share of major devastating geologic events in recent decades, looking back at the historic record it's pretty clear that none of what we have experienced is new or anything different. Floods have occurred with surprising frequency and hit the same areas – communities built on flood plains like Santa Cruz, Soquel, and Pajaro to name a few – again and again. Coastal storms and El Niño events have flooded the same low-lying areas repeatedly, damaging both private development and public facilities – Capitola, Seacliff, Rio del Mar, and those locations where East Cliff Drive dips down to sea level, Twin Lakes, Corcoran Lagoon and Moran Lake. And droughts are nothing new in California, in fact in the not too distant past we have had dry periods lasting for decades.

The big difference today is that Monterey and Santa Cruz Counties are home to 700,000 people, with more on the way. Well that, and the climate is changing. This is going to alter the rules and weather patterns we have grown accustomed to and depend upon, as well as what we may expect in the future with rainfall, floods, droughts and fires.

Many geologists in the past spent their lives breaking open and analyzing rocks and looking for the history they contained in order to try to determine or extract what that particular geologic formation or outcrop told us about what took place 10 or 20 million years ago when the rocks formed. What has been exciting about living, working and teaching in the Monterey Bay region for the past 50 years is that we have actually been able to witness a number of major geologic events and see what history or stories are left behind.

Being the first geologist to hike up the deadly Love Creek Slide on the morning of January 4, 1982, and then mapping the high water marks along the Santa Cruz County streams from that winter's floods, left me with very strong impressions of the power of nature. In October 1989, there were many of us scouring the Santa Cruz Mountains for the cracks and fissures left behind by the intense shaking accompanying the Loma Prieta earthquake to see what we could learn. There was also the pain and suffering caused by the shock that left much of downtown Santa Cruz in ruins, damaged hundreds of homes, and took six lives across Santa Cruz County. The Moss Landing Marine Laboratories were damaged beyond repair, but other than that, Monterey County fared very well, as they usually have during moderate to large earthquakes.

The chapters that follow delve into the natural disaster history of the region, with descriptions and photographs of events as far back as good records exist, usually from the early newspaper files. While property damage during these disasters has been large, and has increased over time as more people have moved to the area and more development has taken place, the actual loss of life – while tragic – has been surprisingly small: only six deaths from the Loma Prieta earthquake, the largest shock to hit the area in 83 years. There has only been a single tsunami death in the area in the entire period of historic records. Floods and large landslides have taken the lives of only about 30 people since 1900. Less than hundred people have died as a result of natural disasters over the past century, less than one person a year.

This is not to say that these natural disasters shouldn't concern us, but rather that it is important to keep them in perspective relative to the risks we face every day: motor vehicle and bicycle accidents – more and more often involving impairment from texting or using phones while using any of these vehicles; drowning; animal bites or stings; falls or fires. I'm also leaving out a few obvious risks that go without saying, incidents involving guns, for example. These activities all present much greater risks to our lives than natural disasters.

You should enjoy all that the Monterey Bay region has to offer, but be cautious and look carefully at where you decide to buy or build a home. My hope is that you find this natural disaster history engaging, interesting, and educational.

CHAPTER 1

Introduction

Although the Portolá expedition couldn't even find Monterey Bay in the fog on their first overland encounter in 1769, thousands of others in the subsequent years have and consider it to be their own piece of paradise. Drawn by the nearly pristine waters of the bay and its marine life, the redwood forested coastal mountains, an ideal climate, and abundant natural resources, the region seems to have everything. Yet the geological processes and weather that have produced this spectacular landscape and have drawn people here for well over a century and a half, have also wreaked havoc on a regular basis, whether floods or droughts, earthquakes or coastal storms. Paradise here comes with a price, and climate change may be raising the ante a bit more.

Two quite different counties, Monterey and Santa Cruz, border Monterey Bay *(Figure 1.1)*. Although fishing, agriculture, and tourism have traditionally been parts of both counties' economies, culturally, socially, and economically, they are as different as night and day. The Monterey Peninsula historically attracted wealthy tourists from the east coast, thanks to the investments and efforts of the Big Four: Stanford, Hopkins, Huntington and Crocker. Today the peninsula has the annual Concours d' Elegance, the 17-Mile Drive, the lush golf courses of Pebble Beach, the trendy and expensive shops and galleries of Carmel, the elegant hotels of Monterey and Cannery Row, and the car races at Laguna Seca.

Santa Cruz, on the other hand, has always attracted a different crowd. Historically the north end of the bay was the vacation or weekend destination for the farm families and other blue collar workers from the Santa Clara Valley and the great Central Valley, usually looking for the cool and often foggy beaches to get away from the valley's heat. It was initially the Santa Cruz Beach Boardwalk, but today, its Woodies on the Wharf, The Cold Water Surfing Classic, the Tin Man Triathlon, and the dirt racetrack at the County fairgrounds, and the city of Santa Cruz is still searching for a major hotel.

Santa Cruz County is the second smallest in the state and arguably has more geologic hazards per square mile than any of California's other 57 counties. Some of the most expensive homes and spectacular views are found along the coastline where beach level homes are regularly battered by waves and bluffs and cliffs continue to retreat, taking streets and houses with them from time to time. Moving inland and up the river valleys through communities like Santa Cruz, Felton, Capitola, Soquel, Aptos, Watsonville and Pajaro, we find repeated flooding as far back in time as the historic records extend. The flat river bottom lands closer to the coast, the areas now covered by Santa Cruz, Watsonville, Castroville and Salinas, are underlain by sand and have high water tables, which makes them highly vulnerable to liquefaction during large earthquakes.

Those who choose to live in the steep terrain of the Santa Cruz Mountains have to contend with winter landslides and mudflows *(Figure 1.2)*. Highways 9 and 17, now major commuter arteries, are hazardous during heavy rains or the occasional snow and ice, and it's not at all uncom-

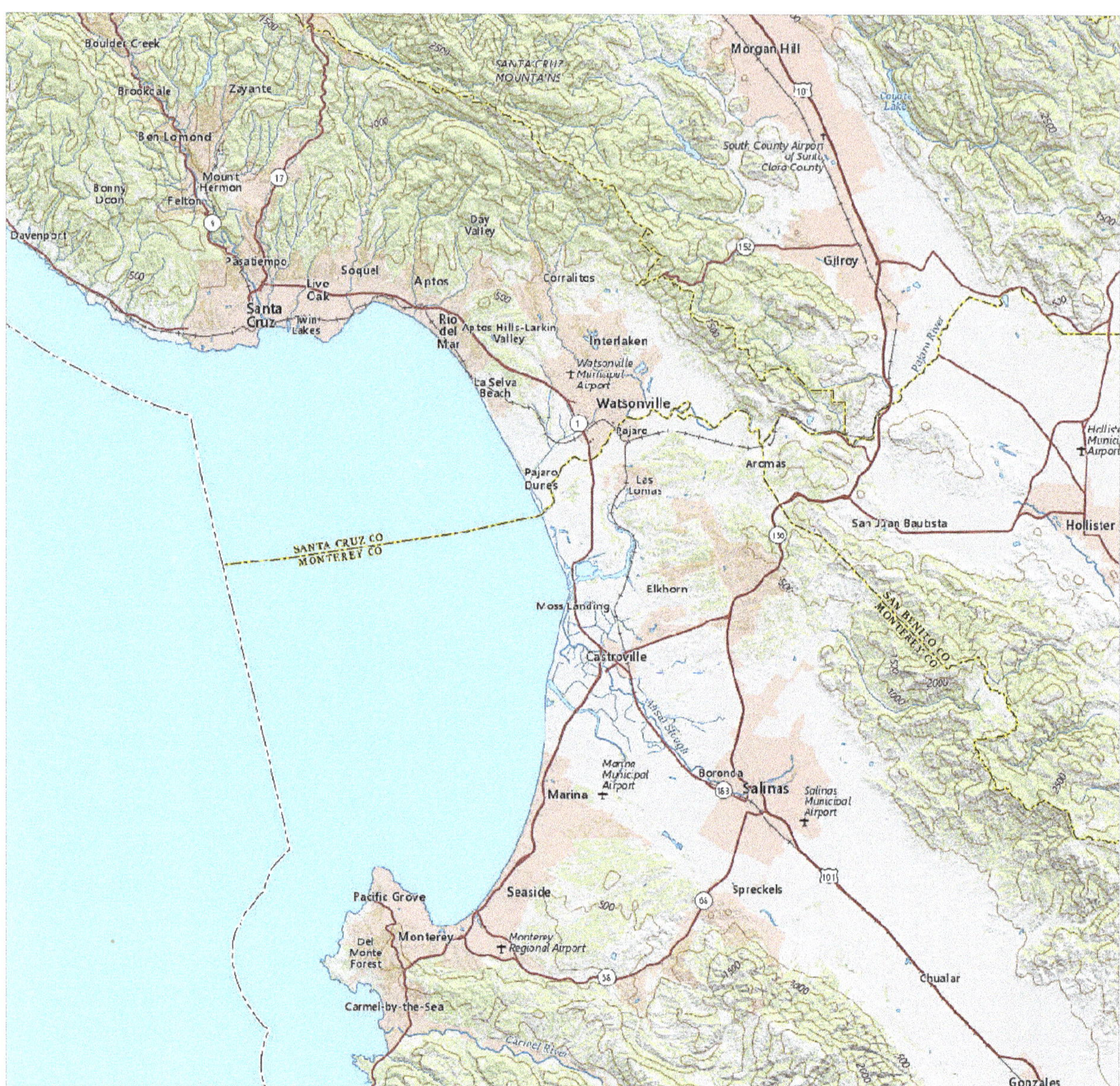

FIGURE 1.1. The Monterey Bay Region. (Map courtesy of the United States Geologic Survey).

mon to drive around one of the many curves on either highway and find a lane closed by rocks and mud. And at the other extreme, the dry years bring high temperatures and an increasing incidence of wildfires to the mountains. Approaching the county line at the summit of the Santa Cruz Mountains, hoping that things will improve, we encounter the San Andreas Fault, the mother of all geologic hazards in California.

Monterey County, by contrast, is one of the state's larger counties, with most of its residents concentrated in a few cities: Monterey, Pacific Grove and Carmel along the coast, and Salinas and other smaller agricultural communities scattered down the Salinas Valley. Much of the county's vast land area, however, lies in the mountains, between Big Sur and the Salinas Valley, and is rugged, nearly inaccessible and only sparsely populated. Nonetheless, California Highway 1 through Big Sur has become a more heavily traveled corridor and drivers have to contend with the occasional fire, often followed by wet winters with landslides, rockfalls, mudflows, subsequent closure and then reconstruction. The 2017 Mud Creek landslide closed the highway for over a year. The Salinas and Pajaro rivers overtop their banks from time to time, inundating and replenishing the surrounding farmland with silt, and occasionally flooding low-lying communities like Pajaro. Droughts or water shortages, however, seem to be increasingly more problematic than floods, particularly on the Monterey Peninsula, where there just isn't enough water to go around. While the Monterey peninsula is underlain by solid granite and about as safe a foundation as you can find, it doesn't hold much water so the cities on the peninsula have had to start looking elsewhere to find water for their residents, visitors, and many golf courses.

FIGURE 1.2. Home in the San Lorenzo Valley demolished by a debris flow during the winter of 1982.

The sandy shoreline of southern Monterey Bay is suffering from long-term retreat with increasing risks to ocean front development. While much of the shoreline around the bay is relatively undeveloped, the southern and northern ends have seen homes built on the cliff tops, on the dunes and on the beach itself. In response to the threats of erosion during stormy winters, the construction of seawalls and revetments was the typical response over the past century *(Figure 1.3)*, but this has changed with more scrutiny now given by the California Coastal Commission to proposals for additional coastal armoring. And a projected increase in the rate of sea-level rise is only going to make a bad situation worse.

Despite all of this geological instability and uncertainty, the regular disasters, storms, floods, droughts, earthquakes and all the rest, people keep coming or coming back here. Any reasonable person would ask, *why?* I think there are several reasons: an important one being that most people tend to have very short disaster memories – sort of a collective amnesia – they just plain forget what it was like in the drought of 1976-77, the floods and mudslides of 1982, the El Niños of 1983 or 1997-98, or the earthquake of 1989. Many who have lived here for decades figure it won't happen again and everything else around the bay is so good, they're just going to take their chances. And before long they just forget about the last flood, earthquake, or drought. And, let's face it, no place is perfect... so why not live in paradise and put up with a few risks and an occasional disaster? Hope for the best but buy insurance for the worst.

Some new arrivals from states east of California, or from across the Pacific, simply don't know any better and read that the Loma Prieta earthquake really took place in Oakland or San Francisco, rather than Aptos. Where's Aptos anyway? And there is an army of real estate agents ready to sell them a piece of paradise during the pleasant days of summer or fall. Perhaps another important reason is that most of the old timers or long-time residents know that there are some recurring natural hazards here, but that the perceived benefits are well worth the gamble, the redwoods, the weather, the coastline and beaches; this place has it all.

FIGURE 1.3. Ocean Harbor House townhouses were built on the front edge of the dunes above Del Monte Beach where erosion rates have averaged 1-2 feet annually. They were originally supported on shallow timber pilings but as a result of continuing erosion and undermining the Coastal Commission approved construction of a concrete seawall.

If you had been born right after the San Francisco earthquake in 1906, you could have lived 83 years, directly on the San Andreas Fault and enjoyed all of the benefits of Santa Cruz County, and never experienced a major destructive earthquake, at least until October 17, 1989. Maybe living with the threat of earthquakes in California is no more dangerous than suffering through tornadoes in Kansas, droughts in Oklahoma, or hurricanes in Florida – hurricanes which became painfully evident in 2017 when Harvey flooded Houston with 50 inches of rain, and Irma rampaged across the Caribbean then smacked both coasts of Florida.

And for most of the residents of Monterey County, the San Andreas Fault heads inland at Chittenden Pass along the Pajaro River and almost misses them completely. So other than some inland damage in 1906 at Speckles and Salinas, ground failure and liquefaction at Castroville and Moss Landing, and damage to the Del Monte Hotel – including the honeymooning couple who were squashed in their bed when a chimney collapsed – life has gone on pretty much as usual without a lot of seismic anxiety.

Most of the geologic and weather-related hazards around Monterey Bay are not at all obvious to the casual tourist or the visitor who decides to buy a beach house or charming mountain cabin in the middle of summer. Its not raining in July; the beaches are wide and sandy, and the mountains are sunny and dry. In the first three months of 1983, however, there were millions of dollars in damage to oceanfront homes and businesses, streets, utilities, and state parks along the shoreline of the bay. Storm waves and high tides – combined with a sea surface elevated by a large El Niño – flooded East Cliff Drive, inundated and battered Capitola, Seacliff, Rio del Mar, Aptos Seascape and Pajaro Dunes. The Capitola Wharf lost its outer end and waves washed completely through the restaurants on the Esplanade. Most of the new timber bulkhead protecting Seacliff State Beach was destroyed for the 7th time. Houses collapsed along Beach Drive in Rio Del Mar; homes on the dunes were undermined at Pajaro Dunes, and the condominium owners in Ocean Harbor House on Del Monte Beach had to install emergency rip-rap to protect their investments.

On the evening of January 4, 1982, after receiving over 15 inches of rain in the previous 28 hours, which came after 35 inches of rain in November and December of 1981, an entire mountainside above Love Creek in Ben Lomond gave way and buried nine homes and ten people. Many San Lorenzo Valley residents have forgotten about Love Creek or never heard of it. Newcomers are just trying to figure out why they named it Love Creek to begin with.

These are not Acts of God or isolated events, and we ought to begin thinking about them just as "*normal*" as any other natural process. The old newspapers and historic records are full of stories of storms and floods, droughts and fires, earthquakes and landslides. Most residents just don't bother to read about local history and others would rather enjoy their bit of paradise without having to worry about either past or possible future calamities. The big difference is that today the Monterey Bay area has a lot more people than it did a century ago. In 1900, the combined population of Monterey and Santa Cruz counties was just over 40,000; by 2018 it exceeded 700,000. Most of the safe, stable and easily buildable parcels were used up decades ago. Early settlers, including the Native Americans, could pick and choose where they wanted to live and they avoided the problem sites. They knew better. Why live right next to the river if you are going to get flooded every other year? So what are left today are often the problem sites, steep hillsides, floodplains, and other sites of questionable long-term stability. Yet if you just drove into the area from Los Angeles, Fresno or Stockton, that redwood-covered "*building site*" looks real good in July. The words "*creekside charmer*" or "*on the sand*" in the real estate jargon sound much more like a ticket to paradise than a recipe for potential disaster to a buyer looking for their piece of heaven.

Santa Cruz and Monterey counties are relatively young geologically speaking. The mountains are still slowly rising and shifting around; in fact the crest of the Santa Cruz Mountains rose about four feet during the 1989 Loma Prieta earthquake. The mountains also moved about six feet to the northwest relative to San Jose during that 6.9 magnitude shock, remembered by many as the precise day and time that the Bay Area World Series between the San Francisco Giants and the Oakland A's was getting underway. The San Andreas Fault represents a major plate boundary, and separates the Monterey Bay area from much of the rest of the country, geologically and culturally. Everything west of the San Andreas Fault is on the Pacific Plate, which is sliding towards Alaska at the rate of about two inches per year. Most of that motion takes place during the infrequent but large earthquakes that periodically release the accumulated strain along this major geological boundary. And about 50,000 commuters drive over the fault – a major plate boundary – every day on Highway 17 on their way to jobs in Silicon Valley.

Santa Cruz County is still rising out of the sea, albeit very slowly, and in fits and starts. Those flat marine terraces or benches, which most of the people of Santa Cruz County and some in Monterey County live on, and which are intensively farmed for Brussels sprouts, artichokes, and occasional pumpkins along the north coast, are testimony to this slow uplift *(Figure 1.4)*. Each of these terraces was formerly at sea level and was eroded flat by wave action over thousands of years. Based on the dating of fossils recovered on these terraces and our knowledge of global sea-level changes, we

FIGURE 1.4. The coastline north of Santa Cruz is characterized by a stair step series of uplifted marine terraces, which provide evidence of continuing tectonic uplift. (Photo: Kenneth and Gabrielle Adelman © 2006).

know that the first terrace (which Highway 1 follows along the north coast of Santa Cruz County on its way to San Francisco) was at sea level about 100,000 years ago. In order to get to its present elevation, it had to have been rising at an average rate of about a hundredth of an inch each year. If you walked at this rate, it would take you about 60,000 years to get across your 50-foot wide backyard. While this may seem inconsequential, over a period of 100,000 years it has given us the real estate that supports about half the population of Santa Cruz County, and much of its agricultural base. It just takes a little time, some forces acting beneath the surface, and we have to be patient.

Heading south from Carmel, the flat portion of Point Lobos and the terraced grazing land you cross on Highway 1 as you approach Point Sur were also formed at sea level about 100,000 years ago and have been slowly raised from an ancient shoreline.

Unfortunately, there are other forces operating to remove this valuable coastal real estate as quickly as it forms, in fact far more quickly. The sea cliffs of Santa Cruz are being eroded at average rates varying from a few inches to a few feet each year. Relative to the hundredth of an inch of annual shoreline uplift, the coastal bluffs are eroding 500 to 2,000 times faster. Put another way, although the coastline is rising, we are not going to see any new ocean front property within our lifetimes; quite the opposite in fact, we will continue to see chunks of coastal bluffs collapse into the rising ocean *(Figure 1.5)*.

While this may be disappointing to the real estate profession, it is a geological fact of life in Santa Cruz County. The Pacific Ocean is 7,000 miles wide and doesn't care too much about a few feet on either edge. Somewhere between 5,000 and 10,000 waves break every day on the sea cliffs of Monterey Bay. The cliffs are going to continue to fail, some areas faster than others. We can shore them up with concrete or protect them with huge rocks to try and reduce the wave impact, but most of our protective efforts should be seen as temporary, especially if the rate of sea-level rise continues to increase as all measurements indicate.

The Monterey County coastline varies geologically far more than the northern end of the bay. The sandy bluffs from the city of Marina to Del Monte Beach are retreating at some of the highest rates in the state. The combined effects of storm wave attack at high tide, gradual sea-level rise, and beach sand mining in Marina, are moving the bluff edge inland at two to six feet every year. This means trouble and an uncertain future for anything built along the dunes of the southern bay shoreline. From Monterey to Carmel Bay, however, the granite is pretty unyielding. Over the past century there hasn't been much change recorded in most places around the peninsula. Granite resists wave attack well. The mere existence of the Monterey Peninsula, jutting out into the Pacific Ocean, is ample evidence for this resistance.

The landscape that draws us and keeps us here, the mountains and sea cliffs, are due to the same awesome geological forces that produce the hazards and disasters, earthquakes, landslides, floods and cliff failure. We're no longer in Kansas, Toto.... this is California.

Know, then, that, on the right hand of the Indies, there is an island called California, very close to the side of the Terrestrial Paradise, and it was peopled by black women, without any man among them, for they lived in the fashion of Amazons. They were of strong and hardy bodies, of ardent courage and great force. Their island was the strongest in all the world, with its steep cliffs and rocky shores. Their arms were all of gold, and so was the harness of the wild beasts, which they tamed and rode. For, in the whole island, there was no metal but gold.

- Las Sergas de Esplandian, 1508

FIGURE 1.5. The 90-foot high cliffs along Depot Hill in Capitola continue to retreat at average rates of about a foot/year, which has led to demolition and relocation of buildings and loss of the cliff top street.

Chapter 2

Earthquakes and Faulting

I feel the Earth move under my feet...
– Carole King

INTRODUCTION

The first ominous rumblings of history's most destructive earthquake since 1556 were heard on a rainy, windblown morning in Beijing, China in July 1976. Two earthquakes, the first of magnitude 8 and a second 16 hours later of magnitude 7, ripped through one of China's most populous regions, crumpling dams and toppling buildings. When it was over the industrial city of Tangshan was totally destroyed, and over 250,000 people had died. The high population density, the typical mud-brick, tile-roofed buildings, and the fact that the first quake struck when most people were indoors sleeping, all contributed to the magnitude of the disaster *(Figure 2.1)*.

The 1976 Tangshan earthquake notwithstanding, from the perspective of earthquake fatalities, the 21st century began ominously and the death and devastation unfortunately continued. In 2001, a magnitude 7.6 earthquake in India led to over 20,000 deaths. Two years later, about 30,000 people perished in Iran in a magnitude 6.6 event. The next year, 2004, was the massive 9.15 earthquake offshore in the Sumatra Trench that produced the greatest fault rupture of any recorded earthquake, spanning a distance of 900 miles, or longer than the entire state of California. Severe shaking lasted for 10 minutes, and the combined fatalities from the earthquake and tsunami were about 230,000 around the northern Indian Ocean.

A year later in 2005, over 87,000 died during a 7.6 magnitude shock in Kashmir, Pakistan, and then a 7.9 magnitude earthquake in Wenchuan, China led to over 80,000 more fatalities. Two years later a large earthquake (magnitude 7.0) struck Haiti, creating near total destruction in unreinforced masonry buildings. There is no agreement on the death toll, which ranges from a low of about 50,000 from a U.S. government report to over 300,000 reported by the government of Haiti, some felt in an effort to obtain more international aid. The most common number of deaths reports is 230,000, but by any measure it was a disaster for Haiti. In striking contrast, and as a testimony to differences in construction methods and building materials, the 1989 Loma Prieta earthquake of essentially the same magnitude (6.9), took 63 lives; a tragedy none-the-less, but on a vastly different scale.

FIGURE 2.1. The library at a university in Tangshan, China, was extensively damaged by a 7.0 magnitude earthquake in 1976 but was left standing as a reminder of the event.

On March 11, 2011, the planet experienced the 2nd massive earthquake of the 21st century, that occurred offshore Japan. The Pacific Plate broke loose and slid beneath the Eurasian Plate producing a 9.0 magnitude shock and devastating tsunami. The official death toll was over 18,000, with another 2,500 people reported missing. Finally, in April of 2015, a 7.8 magnitude earthquake struck Nepal, destroying many ancient buildings and taking over 9,000 lives.

FIGURE 2.2. The ruins of brick buildings in San Francisco following the great 1906 magnitude 7.9 earthquake. (Photo courtesy of the Detroit Publishing Co., Library of Congress Prints and Photographs Division, Washington, D.C.).

And these were only the most devastating of hundreds of other moderate to large earthquakes around the world. But these eight major events alone, led to the combined loss of over 700,000 lives, or an average of 47,000 each year of the 21st century.

Most people in the United States tend to think of the 7.9 magnitude 1906 San Francisco earthquake, in which about 3,000 people died, as a major disaster, and it was for California. Although most of San Francisco was destroyed (due in large part to the fires that followed the earthquake), the number of casualties was relatively low for such a large shock *(Figure 2.2)*. In 1923, an earthquake of similar magnitude near Tokyo killed 143,000 people.

Although California, because of its adolescent geology, contains more natural wonders than most areas of similar size anywhere in the world, it is perhaps best known to those who don't live here for just one thing – earthquakes. California is indeed earthquake country and the accounts, pictures and legends of past earthquakes have left permanent impressions with people across the country and around the world.

Even if the accounts and events accompanying past earthquakes have been exaggerated, California's position in the seismic world, like the economic world, is a very significant one. The state lies on a belt of active faults that circles the Pacific Ocean, and which is responsible for about 80 percent of the world's earthquakes. California's residents experience thousands of earthquakes every year, of which only about 500 are large enough to be felt somewhere. Every four or five years a large, potentially destructive earthquake strikes California, with the greater San Francisco Bay area experiencing about 12 damaging shocks per century, at least during the last 200 years or so of written history. We don't really have significantly more earthquakes than other similar sized areas around the Pacific Rim; but being the nation's most populous state with nearly 40 million people, being Paradise or the Garden of Eden in the minds of many who don't live here, and being home to so many immigrants, earthquakes here often seem to get more attention across the country and around the world than those in many other places.

Monterey Bay's seismic history parallels that of the rest of California. Since the mission days of the late 1700s, the occurrence of earthquakes of varying intensity has been a regular and recurring part of the regional history. The central coast has been seismically active for millions of years and will continue to experience earthquakes for millions of years to come. It's a geological fact of life for the region and there is absolutely nothing we can do to change this. We cannot even predict when the next one might happen, and may never be able to.

Before giving up all hope, however, it is important to remember that all earthquakes are not created equal. The small and moderate events that rattle our windows, doors and chandeliers are relatively common and painless, but serve as monthly reminders that we live in earthquake country and that the large tectonic plates we live on are constantly shifting and adjusting. These minor disturbances give us pause to think about what we will do when the *"big one"* takes place, and then most of us calmly forget about it and go back to our computer screens, iPads or iPhones. We nearly always have more pressing or immediate things to worry about.

Unfortunately there won't be a single *"big one"* in California; there have been many big ones and there will be many more. Not frequently, but they will occur. They have to for a very simple reason – California straddles the boundary between what we call the North American Plate and the Pacific Plate. In many of the world's earthquake-prone regions, the plate boundaries, fault lines and fractures are largely invisible, many lying deep beneath the ocean surface. In California, however, the boundary between these massive plates is called the San Andreas Fault and it is visible from the air and ground in many places *(Figure 2.3)*. It is a ragged scar, a zipper, crossing the landscape for 700 miles from the Gulf of California to Cape Mendocino, north of San Francisco, where it dives off into the Pacific. The San Andreas and its branches slice through farms and forests, through cities, under reservoirs, freeways and universities. Over 20 million people live near this massive break in the earth's crust and its associated fractures. Their lives and property are all affected and will continue to be.

FIGURE 2.3. The San Andreas Fault crossing the Carrizo Plain in central California has produced a very visible gash across the landscape. (Photo: David K. Lynch © 2018, www.sanandreasfault.org).

Just east of Monterey Bay, a major offshoot of the San Andreas, the Calaveras Fault, is slowly tearing the small rural town of Hollister apart. It's well worth the drive to this tidy little farm town on a Saturday or Sunday when everyone else is driving the other way to the beach in Santa Cruz. You don't have to be a geologist to appreciate that the streets, curbs, fences and houses are gradually being pulled apart as the opposite sides of the fault creep in different directions *(Figure 2.4)*. The city keeps patching up the streets and sidewalks, and homeowners annually repair their chimneys and repaint their houses. But the hidden forces just beneath the ground continue stretching Hollister like a rubber band. And the residents go right on patching and painting.

FIGURE 2.4. Creep along the Calaveras Fault as it passes through the small farming town of Hollister has gradually offset streets and sidewalk, fences and walls, as well as houses.

Heading north, the San Andreas leaves San Benito County and slips quietly into Santa Cruz County from the southeast at Chittenden Pass (Highway 129), having never so much as touched Monterey County. The course of the Pajaro River is deflected by the fault as it winds its way towards Monterey Bay. In the 1906 earthquake the railroad bridge crossing the river was badly damaged when the concrete piers were cracked and moved a few feet towards the river *(Figure 2.5)*. Unfortunately, a freight train was rolling along south of the bridge at about 30 miles an hour at the time of the earthquake. To the surprise of the engineer, ten cars in the middle of the train were thrown off the track.

FIGURE 2.5. Severe shaking and settlement of river bottom sediments during the 1906 San Francisco earthquake led to the partial failure of this railroad bridge crossing the Pajaro River in Chittenden Pass. (Photo courtesy of the Carnegie Commission Report, 1908).

At the right angle bend in Highway 129 in the middle of Chittenden Pass, you can almost put your hand on the San Andreas Fault, one of the few places in California where this is possible. If you decide to try this, however, make sure you watch out for the large trucks that barrel along this road completely oblivious to the fault's presence. The shale on the steep cliffs on the east side of the highway is separated from the granite in the huge Logan Quarry on the opposite side of the river by the fault. The granite is a chunk of the Sierra Nevada that has been carried northward along the fault from southern California over the past 20 million years. Other large and displaced masses of granite from the Sierra Nevada make up a part of Pt. Lobos, the Monterey Peninsula, the Santa Cruz Mountains, the Farallon Islands and Bodega Head.

Northwest of Chittenden and behind Watsonville, the path of the fault follows the western flank of the Santa Cruz Mountains, offsetting the wooded stream valleys that drain the steep grassy hillsides. Countless earthquakes over thousands of years have produced a hummocky landscape, much like a rumpled carpet, characterized by landslides and slumps, linear ridges, sag ponds and springs *(Figure 2.6)*. This area can best be seen from Carleton Road between Hecker Pass (Highway 152) and Chittenden Pass (Highway 129). On a clear day, the fault zone topography is also visible driving by Watsonville on Highway 1 as a bench halfway down the grassy mountainside *(Figure 2.7)*. Thousands of years of earthquakes have broken up and weakened the underlying sandstone and shale and disrupted both the surface runoff and ground water flow. The impoundment of the water results in a number of small ponds and lakes, not visible from the roads below, which dot the landscape and topographic maps. For the most part, this area is uninhabited, however, and only the grazing cattle and a few scattered residents feel the small but frequent earthquakes.

Hearing northwest from Hecker Pass, the fault crosses both Hazel Dell and Mt. Madonna roads as it continues off into the redwood-covered slopes of the Santa Cruz Mountains. As the grasslands turn to forest the somewhat anomalous and confused topography created along the fault is more difficult to recognize. Further to

FIGURE 2.6. Aerial view of the San Andreas Fault Zone on the western flank of the Santa Cruz Mountains east of Watsonville. Thousands of years of fault movement have weakened the underlying sedimentary rocks such that landslides and slumps form the dominant features of the landscape.

FIGURE 2.7. The trace of the San Andreas northwest of Pajaro Gap is clearly delineated by a valley along the fault and stream courses, many of which are displaced to the left as they cross the fault. (Image: Landsat / Copernicus via Google Earth).

the northwest, Eureka Canyon Road and Highland Way follow the path of the fault for several miles. Buzzard Lagoon is a sag pond along the fault trace. The upper reaches of both Corralitos Creek and the East Branch of Soquel Creek also flow along the fault valley. Few people live in this rugged area although there are some schools and summer camps scattered throughout the fault zone and this portion of the southern Santa Cruz Mountains.

The southwestern flank of the Santa Cruz Mountains, all the way to Highway 17, was the scene of many large landslides in both the 1906 and also the 1989 earthquakes, and the few mountain homes existing at the time of the early quake suffered major damage. Mountain home development along the Summit has proliferated in recent years, due primarily to easy access to Highway 17 and Silicon Valley. Home damage during the 1989 quake from both seismic shaking and also ground cracking was widespread. Many large older slides were reactivated, but only moved a few feet due to the very dry conditions preceding the October 17 event.

The San Andreas Fault follows Laurel Creek for several miles, passes through the Summit Road/Old San Jose Road intersection and then under the Loma Prieta School. Extensive cracking and offset roads and fences were common in this area after the 1906 shock. At the Morrell Ranch, a mile south of Wrights on the Summit, a crack opened up under the house, and tore it in half *(Figure 2.8)*. Lateral offset here reached six feet. Although rupture along the fault itself in the Summit Ridge area in 1989 was not as extensive or clear as it was in 1906, ground cracking from shaking and sliding were widespread and produced damage and destruction throughout the Summit area *(Figure 2.9)*.

FIGURE 2.8. The Morrell House along the crest of the Santa Cruz Mountains near Highway 17 was severely damaged by strong seismic shaking during the 1906 San Francisco earthquake. (Photo courtesy of the United States Geological Survey).

FIGURE 2.9. Extensive ground cracking took place along the summit area of the Santa Cruz Mountains during the 1989 Loma Prieta earthquake.

Heading northwest, Santa Cruz County is technically free of the San Andreas Fault as it dives off into Santa Clara County and follows the north side of Summit Ridge. A 6,200-foot long tunnel along the old San Jose-Santa Cruz Railroad connected Wrights in Santa Clara County with Laurel in Santa Cruz County at the turn of the last century. In 1906, however, the fault ruptured the tunnel 400 feet in from the Wrights entrance and offset the tunnel and tracks five feet. The tunnel was never reopened and this spelled the end of the Sunshine Special rail connection that brought summer visitors through the mountains to the beaches of Santa Cruz a century ago.

WHY DO WE HAVE EARTHQUAKES?

Over the past 50 years or so, geologists and geophysicists have developed a much clearer understanding of why earthquakes occur where they do. A glance at a map of worldwide seismicity quickly reveals that patterns are not random, but that most earthquakes are concentrated along some very narrow zones, one of which follows the

San Andreas Fault, for almost the entire length of California *(Figure 2.10)*.

There is now good agreement among Earth scientists that the Earth's crust (or lithosphere) is broken up into a number of large, rigid plates about 60 miles thick that are moving around relative to one another due to the motion of hot fluid material within the underlying upper mantle (asthenosphere) of the Earth. Earthquakes are concentrated at the edges of these massive plates where they interact. Some of these plates are coming apart along a 25,000 mile long volcanic mountain chain, which splits the middle of the Atlantic and Indian oceans and crosses the eastern Pacific, looking a lot like the seam on a baseball – a very large baseball *(See Figure 2.10)*.

Earthquakes occur along these seafloor fissures where hot molten material forces its way up from the mantle to erupt and form undersea volcanoes. Thousands of miles away, the opposite sides of these plates form a second type of plate boundary as they collide. Something has to give at these head on collisions, and deep trenches form where thin, dense oceanic plates are forced down beneath lighter and thicker continental plates. As the plates scrape against each other, large earthquakes frequently occur. Collision plate boundaries of this type occur almost completely around the margins of the Pacific Ocean, which is marked by an almost continuous series of deep narrow trenches. The Earth's largest earthquakes take place at these subduction zones – offshore and underneath Alaska, Japan, the Philippines, New Zealand, Chile, Peru, Costa Rica and Mexico.

The San Andreas Fault is a third type of boundary, where two huge plates grind continuously but very slowly alongside one another. Similar faults occur in New Zealand and Turkey. The San Andreas Fault zone, with its associated branches and splinters, is 50 miles wide in the Monterey Bay area. This zone separates the North American Plate, which extends from the summit of Highway 17 all the way to the middle of the Atlantic

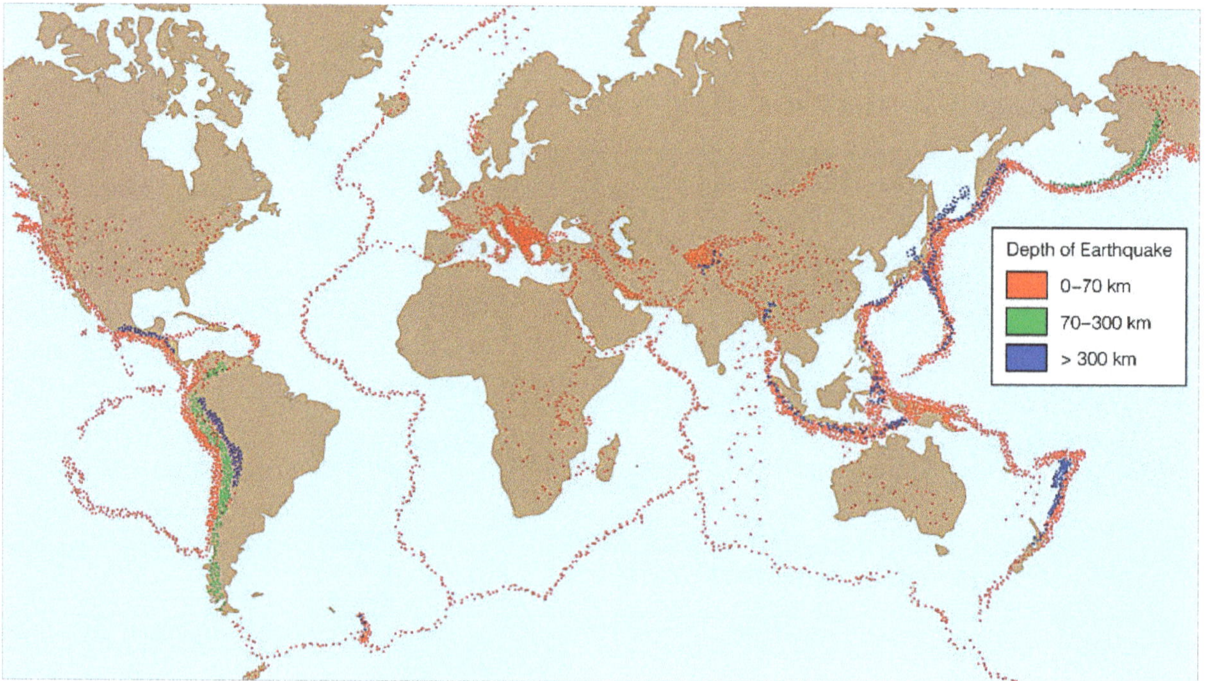

FIGURE 2.10. The global distribution of earthquakes outlines the boundaries of the Earth's tectonic plates. (Illustration courtesy of the University of California Press).

Ocean, from the Pacific Plate, which consists of almost half of the earth's surface, the entire Pacific Ocean extending clear to Japan *(Figure 2.11)*. This is a huge piece of real estate. And these two plates just happen to rub against each other right here in our backyard. They are moving alongside each other at about two inches a year. There isn't a smooth lubricated surface between them, however, so they tend to stick together until decades of strain accumulates, the rocks finally rupture and the plates move. That break in the rocks initiates an earthquake, which sends seismic waves out in all directions at several miles per second. This unfortunately makes calling someone to warn them of an impending quake after you felt it, of somewhat limited value.

If we accumulate two inches of strain each year along the San Andreas Fault, in 50 years we would have the potential for an earthquake that could produce up to 8 feet of horizontal displacement; in 100 years, the accumulating strain could now generate over 16 feet of rupture, and so on. Based on the rate of movement along the fault, the approximately 200-year historic earthquake record, and excavations into the soils and sediments along the fault where we can recognize evidence for large prehistoric earthquakes, we now know that large earthquakes (magnitude 7 or greater) occur about every 50 to 100 years along the San Andreas Fault system. The entire fault doesn't rupture at once, however, and individual segments of the fault can behave differently.

It is now believed by the experts (a respected group of scientists known as the Working Group on California Earthquake Probabilities) that the segment of the San Andreas Fault passing through Santa Cruz County and extending up through San Francisco has about a 22% probability of generating a magnitude 6.7 earthquake or larger in the 30 years from 2014 to 2043 *(Figure 2.12)*.

The Parkfield segment of the fault, passing through the southern end of Monterey County, about 75 miles southeast of Monterey Bay, had

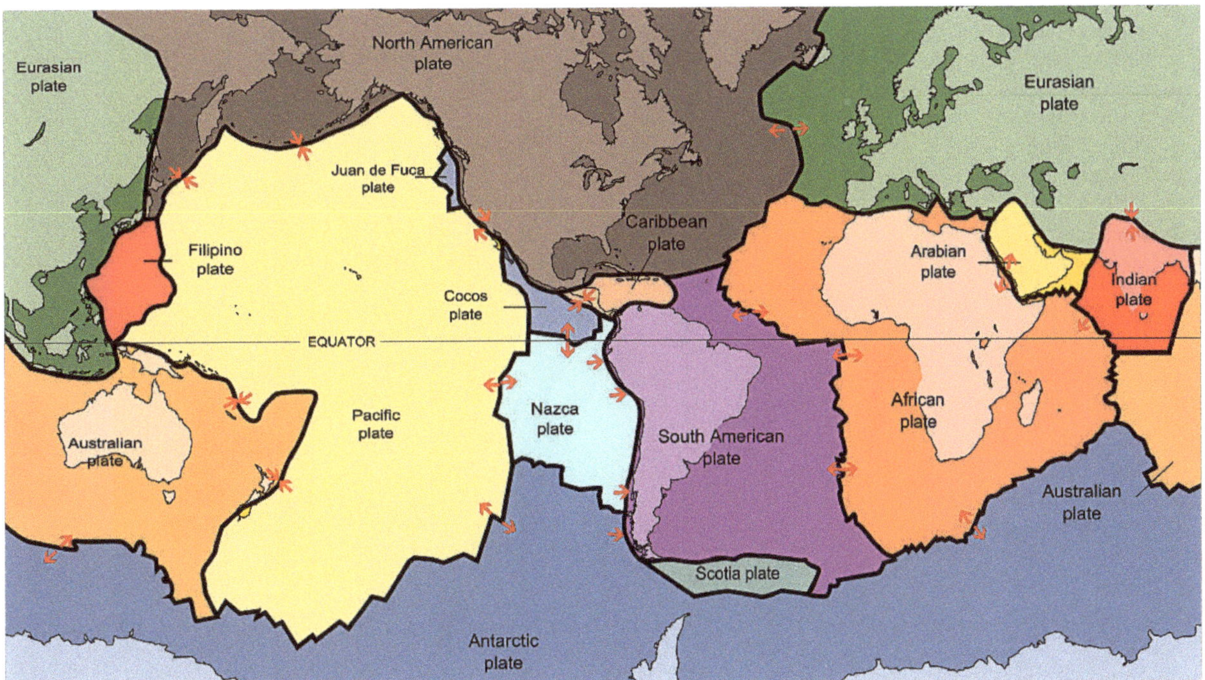

FIGURE 2.11. The Earth's major tectonic plates showing relative motion at plate boundaries. (Illustration courtesy of the United States Geological Survey).

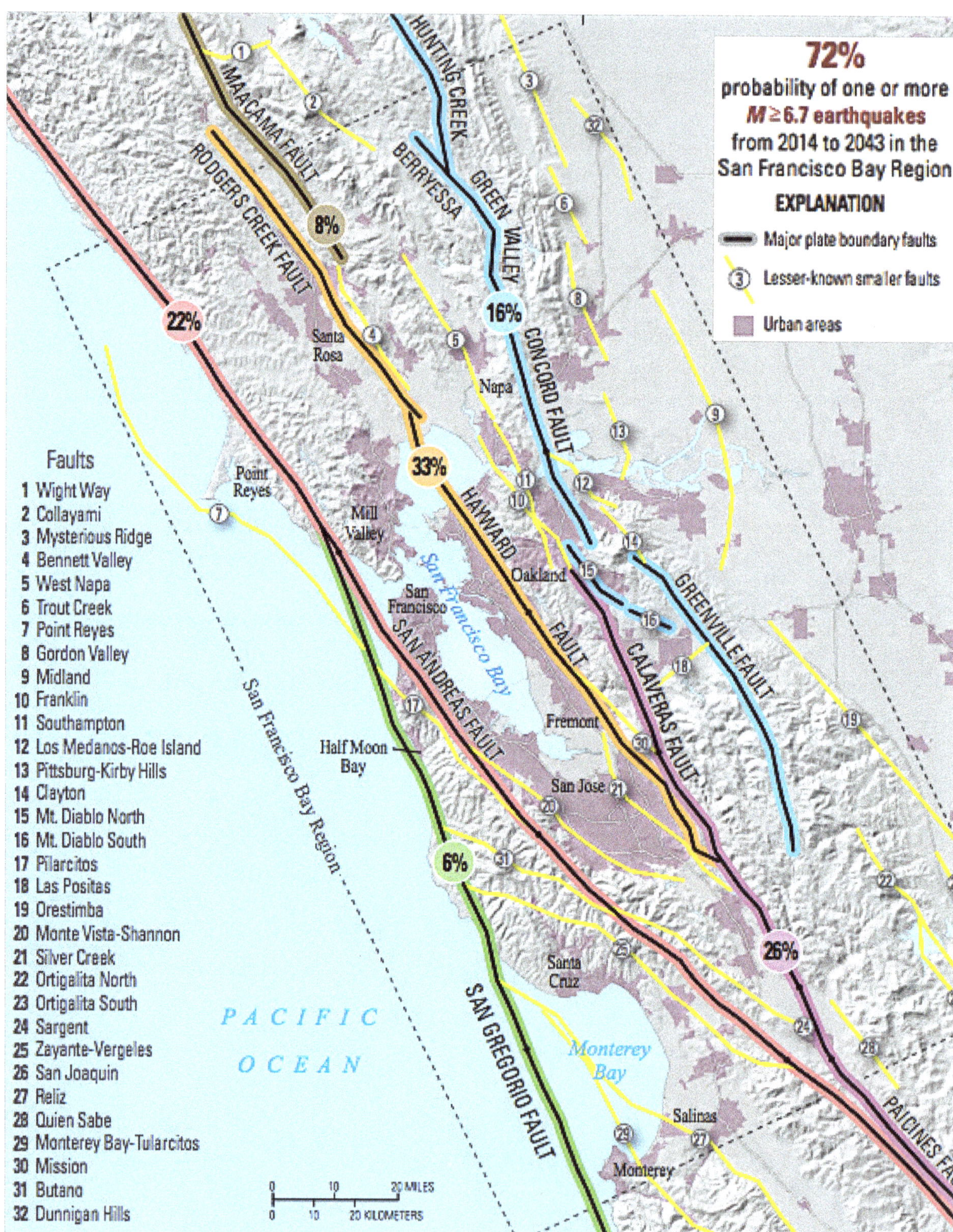

FIGURE 2.12. Major faults in the San Francisco Bay area showing a combined 63% probability of one or more magnitude 6.7 or larger earthquakes occurring between 2014 and 2043. Probabilities of magnitude 6.7 or larger earthquakes on individual faults are labeled on those faults. (Illustration courtesy of the United States Geological Survey).

slipped every 20-25 years throughout the past century producing earthquakes in the 6.5 magnitude range. Because the last earthquake had shaken the Parkfield area in 1966, scientists felt the next one was likely about 1986 and decided to place an array of instruments along his part of the fault. Their objective was to record any possible precursors to the next quake so as to get a clearer idea of whether any measurements or observations might prove useful for predicting future earthquakes elsewhere. Despite the historical regularity and all of the instruments that had been installed, the pattern didn't continue and it wasn't until September 28, 2004, 38 years after the previous quake when a magnitude 6.0 earthquake finally rocked Parkfield. Earthquake prediction has proven to be almost completely elusive so far, not only in California, but also globally.

WHAT HAPPENS DURING AN EARTHQUAKE?

Although many early civilizations, as well as our own Native Americans, had their own beliefs and legends about the causes of earthquakes, it wasn't until the great San Francisco event of 1906 that we first understood the significance of the San Andreas Fault. That event showed us that it is breakage or displacement along a fault that produces the intense ground shaking we feel and that can destroy buildings and claim lives. As the rocks rupture or break, the strain that has accumulated for decades is released as seismic waves. These waves radiate out away from the zone of rupture, much like the ripples that radiate outward from a pebble thrown into a pond. There are several different kinds of waves that are generated, but they all shake the ground and they all can cause damage; the stronger and longer the shaking, the greater the damage is likely to be.

In general, the size of the earthquake, the distance from the rupture zone, the type of rock or soil we build on, the materials we build with, and the quality of construction are the most important factors that affect the damage to be expected during a quake in any particular location. Everything else being equal, which is rarely the case, the larger the earthquake and the closer we are to it, the greater the shaking and damage we can expect.

As we move from hard crystalline rock like granite, to softer sedimentary rocks like sandstone or shale, to loose sediment like silts and clays, and finally to water saturated materials like river bottom or shoreline sediments, we can expect the damage during an earthquake to substantially increase. Unfortunately it is the flat alluvial plains along our rivers that many of the communities around Monterey Bay were built on, Santa Cruz, Watsonville, Pajaro, Castroville, and Salinas being unfortunate examples. The downtown areas of these communities have experienced much greater damage than other areas during past earthquakes because of the loose sediments and high water tables beneath them. Compared to bedrock, these areas behave more like jello during an earthquake. The Monterey peninsula, in contrast, has generally behaved pretty well, with limited earthquake damage. Granite provides a solid foundation and doesn't shake as violently as loose sand and mud.

Much of the historic development around the margins of San Francisco Bay was sited on unstable ground, often artificial fill, which became painfully clear in both the 1906 and 1989 earthquakes. Reports on the great San Francisco earthquake documented the exaggerated shaking and consistently greater damage to buildings in the lower waterfront areas of the city, which were underlain by the thickest bay mud and land fill, as compared to those buildings in the higher bedrock hills. Moreover, the nearby cities of Salinas, San Jose and Palo Alto, built on deep alluvial soils, suffered to an extent far out of proportion to their distance from the epicenter in 1906. Again, in 1989, structural damage in San Francisco's

Marina district and the tragic collapse of the Cypress freeway overpass in Oakland, which was responsible for two-thirds of the deaths in the earthquake, took place 75 miles from the epicenter and was due to the amplified seismic shaking in the weak underlying sediments *(Figure 2.13)*.

FIGURE 2.13. Collapse of the Cypress Street Viaduct along the waterfront area of the East Bay during the 1989 Loma Prieta Earthquake due to the severe shaking of the weak foundation materials led to the deaths of 42 people, 2/3 of the overall deaths during the earthquake. (Photo courtesy of Howard Wilshire, United States Geological Survey).

While some very large cracks opened up in the Santa Cruz Mountains during the 1989 Loma Prieta earthquake, these don't usually swallow up people and animals. Ground cracking is a common earthquake effect, whether due to movement along the fault itself, or to landsliding, settling, or liquefaction. Maximum horizontal offset along the San Andreas Fault in 1906 reached about 15 feet near Pt. Reyes, north of San Francisco. In Santa Cruz County the greatest offset documented was about 6 feet in the old railroad tunnel beneath the Santa Cruz Mountains. While displacement of this sort along a fault is usually confined to a relatively narrow zone, when the fault ruptures there is little we can do to control it, other than by completely avoiding the fault zone to begin with. Because fault zones may be hundreds of feet wide and don't always break in the same place, it is difficult, even for geologists, to know precisely where the fault may rupture next.

Most scientists were very perplexed that the 1989 6.9 magnitude earthquake, despite all of the death and destruction that it produced, did not result in any significant or easily recognizable fault rupture at the ground surface. While there were many cracks and fissures resulting from the earthquake, particularly in the Summit Ridge area of the mountains *(See Figure 2.9)*, most of these were believed to be due to the activation and slight downslope movement of some large older landslides, or simply due to the extreme shaking along ridge tops, rather than movement along the fault itself. To the typical homeowner who has had their house torn in half or moved off its foundation *(Figure 2.14)*, the reason or source for the crack is of little importance. They lost their house and they're understandably upset. The lesson to be learned, which is often forgotten, is not to rebuild over the old cracks. Covering the cracks with dirt helps our collective amnesia, and may make it easier to sell the property to an unsuspecting buyer, but doesn't change what is happening beneath the ground.

What is important to understand if we are to reduce damage and destruction from future earthquakes is that we need to learn from the disasters of the past. The same kinds of processes tend to affect the same areas repeatedly. The detailed reports from the early geologists who traversed Santa Cruz County and the Monterey Bay area on foot and horseback after the 1906 earthquake sounded strikingly familiar to the newspaper accounts and geologic investigations of what we witnessed in 1989. While we cannot predict precisely when future earthquakes will occur, we do have a clear idea of where they will occur and how specific areas will be affected. This isn't a mystery and it isn't rocket science, its just learning from the past.

FIGURE 2.14. House in the Summit area of the Santa Cruz Mountains where strong seismic shaking led to floor joists being ripped off of the mudsill, as well as failure of portions of the mudsill. Seismic tie downs were not required when this home was built.

EARTHQUAKE HISTORY OF THE MONTEREY BAY REGION

While hundreds of small to moderate earthquakes are felt each year in the Monterey Bay region, what follows is a description of the larger historic shocks as reported by various sources, including the local newspapers. Not surprisingly, the further we go back in time, the greater the uncertainty and the less information is available. Seismographs did not come into common use until the early 1900s, and were followed by the development of the first magnitude scale by Charles Richter in 1935. Prior to this time, earthquakes were described by the intensity of shaking as observed by people who experienced the earthquake firsthand by using the Modified Mercalli Intensity Scale *(Table 1)*. The effects of a single earthquake will vary widely from place to place depending upon the distance to the epicenter and the surface materials, but an intensity of VI or higher is indicative of a moderate to large earthquake with a high likelihood of some damage.

Intensity	Description of Shaking / Damage
I	Not felt, except by very few.
II	Felt by person at rest, those on upper floors, or those favorably placed.
III	Felt indoors. Vibration is like the passing of light trucks.
IV	Vibration is like the passing of heavy trucks.
V	Felt outdoors. Small unstable objects are displaced or upset.
VI	Felt by all. Furniture is moved. Weak plaster and masonry cracks.
VII	Difficult to stand. Damage occurs to masonry and chimneys.
VIII	Partial collapse of masonry. Frame houses are moved.
IX	Masonry structures are seriously damaged or destroyed.
X	Many buildings and bridges are destroyed.
XI	Rails bent greatly and pipelines are severely damaged.
XII	Damage is nearly total.

TABLE 2.1. Modified Mercalli Intensity Scale.

1825

An historical account of Mission Santa Cruz written by Paul Johnson states that.... *"a series of earthquakes....caused so much damage that it took years to repair the buildings."*

JUNE 1838

Little has apparently been recorded about this event but what information is available has recently been compared to more recent earthquakes. The conclusion drawn is that this shock probably ruptured the Loma Prieta segment of the San Andreas Fault as well as the segment to the north extending perhaps to San Francisco. The magnitude of the earthquake is estimated to have been about 7.2.

JANUARY 16-18, 1840

Four different historic sources state than an earthquake destroyed the Santa Cruz Mission and a "*tidal wave*" carried many tiles to the sea. This story, however, has been discounted by the Santa Cruz Historical Society, which stated in October of 1973:

"one of the bells cracked on its initial sound and one broke when the tower collapsed in 1840. No rain or earthquake happened on that date. No excuse except poor material or workmanship".

There is no record of this earthquake in the history of Villa Branciforte, although it is thought that a severe storm did occur at this time. The Santa Cruz Mission was built on the first marine terrace about 75 feet above sea level so there is no possibility that it has ever been reached by any waves from the ocean, tsunami or otherwise. As is often the case with older events, one incorrect account is written down or published and later writers or historians simply repeat the mistake, and might even embellish the story.

FEBRUARY 26, 1864

Southern Santa Cruz Mountains with a strong shock in Santa Cruz (Intensity VI).

DECEMBER 18, 1864

Strong shock in Watsonville and Santa Cruz (Intensity V).

OCTOBER 8, 1865

"The most severe shock since the annexation of this territory" occurred on this date in 1865. Recent re-analysis indicates that the earthquake was probably centered in the Santa Clara Valley and the eastern foothills of the Santa Cruz Mountains, and perhaps not on the San Andreas Fault. It is thought to have been one of the five largest quakes to strike the San Francisco Bay region in historic times. The magnitude of this quake has recently been estimated as 6.5, so a little smaller than the Loma Prieta event. The earthquake caused damage from San Juan Bautista on the south to Napa on the north, and ruined the city hall in San Francisco as well as damaging water and gas pipes. Monterey escaped unharmed, however.

The first reports from Santa Cruz stated "*every brick building here is ruined*". Total losses in the city were estimated at $10,000 (this statement suggests that the number and worth of the city's brick buildings was not terribly high in 1865). The ground settled along the San Lorenzo River, cracking the soils along its banks. Some small cracks emitted jets of water two to four feet high for several minutes indicating liquefaction in the subsurface. Near Soquel, the sea was reported to be rising and falling with "*convulsive throbs,*" carrying some of the high cliffs into the sea (which was presumably in the vicinity of Capitola at that time).

In Watsonville, there was a grand total of $2,000 in damage, $1,500 of which was at the Pajaro Flouring Mills. It was stated that every merchant in town lost between $10 and $150 in crockery and glassware. Well-constructed buildings or those built on solid ground suffered little or no damage. Cracks appeared on the banks of the Pajaro River ranging from 10-15 inches wide and hundreds of yards long. Later, observers remarked "*it is a singular fact that the shock was most severe at Santa Cruz and along the lower part of the Pajaro River.*" This statement is a clear indication that residents noticed 150 years ago that the river bottom land with deep sediments and a high water table (like downtown Castroville, Watsonville and Santa Cruz) would experience more intense seismic shaking than areas on solid ground or bedrock. These same patterns have been repeated in virtually every subsequent large earthquake in central California.

Highest intensities during this event were felt in the mountains between Santa Cruz and San Jose. At Mountain Charlie's on the Santa Cruz Road, the earth opened up in several places and steam and water were emitted from the cracks. On the Santa Cruz Gap Road, chimneys were thrown down and the roads were more or less obstructed by boulders that rolled from the hillsides. Wells and streams in Santa Cruz County were markedly affected as many of their volumes doubled. Water also boiled up from the ground for half an hour after the shock. This was observed at the old Santa Cruz Mission orchard, which was presumably down on the flood plain.

OCTOBER 21, 1868

One hundred and fifty years ago, in the early morning of October 21, 1868, an earthquake with an estimated magnitude of 6.8 shook the greater San Francisco Bay area for about 40 seconds. This earthquake was the largest since 1776 when record keeping was initiated, and until the great 1906 event, was known widely as the San Francisco Earthquake. Surface rupture was recorded along about 20 miles of the Hayward Fault on the east side of San Francisco Bay where damage was extensive *(Figure 2.15)*. Six feet of horizontal displacement was reported to have taken place and although population was low at the time, losses

FIGURE 2.15. The Alameda County Courthouse suffered major damage during the 1868 earthquake along the Hayward-Calaveras Fault System. (Photo courtesy of the University of California Berkeley Seismological Laboratory).

were great with almost every building in the small town of Hayward damaged to one degree or another. Damage also occurred in Oakland, San Francisco and San Jose with a loss of thirty lives.

Trenching into the subsurface along the Hayward Fault in recent years has led to the recognition of evidence for 12 large earthquakes. Five of these occurred between 1315 and 1868, with an average recurrence interval of 138 years. With the last major earthquake having taken place 150 years ago, U.S. Geological Survey scientists believe the next one could occur at any time.

While the 1868 epicenter was about 50 miles away from the Monterey Bay area, shaking was still intense. Brick buildings cracked in Santa Cruz but none toppled (Intensity VI). Chimneys that fell were often at the same locations as those that fell in 1865 (apparently brick chimneys were rebuilt the same way, probably without reinforcing steel at that time). The old Courthouse (Cooper House) experienced some cracking of plaster. In Soquel a few chimneys were dislocated and plaster cracked in Watsonville. A 50-foot wide debris flow at Eagle Glen (now known as Majors Creek) carried rocks and trees 1,000 feet downslope into the canyon below.

APRIL 24, 1890

An earthquake of estimated magnitude 6.3 shook the Monterey Bay region and was believed centered near Pajaro Gap; although damage was slight in Santa Cruz many chimneys were thrown down in Watsonville. Ground fissures opened up in the San Andreas Fault zone near Chittenden Pass and the railroad bridge over the Pajaro River was displaced 18 inches. Landslides from the steep bluffs closed the railroad and highway. At Glenwood the ground cracked, and in Boulder Creek *"babies were rolled out of their cradles, clocks were stopped, and any amount of dishes were broken"*.

APRIL 18, 1906: SAN FRANCISCO
THE GREAT EARTHQUAKE (AND FIRE)

The magnitude 7.9 1906 San Francisco earthquake, which unzipped 270 miles of the San Andreas Fault from Shelter Cove, near Cape Mendocino, to San Juan Bautista, east of Monterey Bay, was the most destructive to central California in historic time. Although the epicenter is now believed to have been just off the Golden Gate, the shaking was felt from Coos Bay, Oregon to Los Angeles, an area of 350,000 square miles. The 1908 Carnegie Commission report provided a very thorough investigation of the effects of the earthquake throughout California by a group of dedicated geologists who traversed the region, primarily on horseback, to document the extent of the ground failure and damage. Some selected observations are included below, in some cases, exactly as written (in italics). This was the event that made all of California, including its geologists, aware of the role and significance of the San Andreas Fault. The impacts of the shaking from this shock turn out to be good predictors of where damage would occur in future large earthquakes, such as Loma Prieta in 1989. Where communities were built on deep flood plain soils (downtown Santa Cruz, Capitola, Soquel and Watsonville, for example), damage was far greater than the terraced areas underlain by bedrock (the *"low ground"* vs. the *"high ground"*).

SANTA CRUZ COUNTY

Santa Cruz

"The city of Santa Cruz furnishes excellent evidence on the effect of soil formation on the intensity of the earthquake shock. On the high ground in Garfield Park, and also in the northwest part of the city, only about one-fourth of the chimneys fell and a little plastering was cracked; while in the lower ground near the business section several brick and stone buildings were partly shaken down. The San Lorenzo River was churned into foam, the banks cracking and settling several inches; and sand, said to have come from a depth of

100 feet, was forced up in several places. The bed of the river is also said to have sunk several inches..."

The courthouse (the old Cooper House on Pacific Avenue – *Figure 2.16*) was almost destroyed as the cupola fell through the ceiling and landed in the basement *(Figure 2.17)*. Plate glass windows broke along Pacific Avenue and at least one brick building collapsed. It was estimated that one third of the chimneys in the city were either destroyed or damaged [Does this sound like the 1989 Loma Prieta earthquake?] and all of the bridges across the river were reported as badly wrenched and declared unsafe. Cracks opened up in the street near the railroad depot and at the corner of Front Street and Soquel Avenue. The road near the old Riverside Hotel dropped several inches. Broken water mains and the 8-inch city water pipe at Wilder's Dairy were broken and twisted, which shut off the water supply. All telegraph lines between Santa Cruz and points north were thrown down keeping the city uninformed of the destruction of San Francisco for two days. Mail service was cut off for four days isolating the city and leading to many rumors. One that circulated in San Jose spoke of Santa Cruz being carried into the sea by a tsunami.

FIGURE 2.16. The former Santa Cruz County Courthouse (Cooper House). (Photo courtesy of the University of California Santa Cruz Special Collections).

FIGURE 2.17. Damage to the Santa Cruz County Courthouse (Cooper House) following the 1906 San Francisco earthquake. (Photo courtesy of the Carnegie Commission Report, 1908).

Capitola

"Nearly all of the chimneys at Capitola fell, and considerable plaster was shaken from the north walls of the first floor of the hotel.... Much earth fell from the bluffs near the town...but there was no appreciable effect on the surf. At the county bridge across Soquel Creek, the ground at the east abutment moved inward, cracking the concrete and buckling a water pipe...a continuous cloud of dust rose along the cliffs between Castro's Landing (now called Rio Del Mar) *and Santa Cruz..."*

Soquel

"In the low ground at Soquel, nearly all of the chimneys fell, but most of those on high ground stood. Much plaster fell and goods were thrown from shelves in the business section, which is close to the creek.... Through Delmar, Seabright, and Twin Lakes, nearly all of the chimneys were either down or twisted part way around and left standing, an unusual number being thus twisted..."

Watsonville

Watsonville is built over the loose alluvium of the Pajaro River Valley, where seismic shaking is amplified. The city suffered even more damage than Santa Cruz: *"about 90 per cent of the chimneys were broken off at the roof-line, the greater portion being near to the river... Parts of a few brick walls near the river fell, and considerable settling of the ground took place in Chinatown on the southern side of the river." (Figure 2.18).*

FIGURE: 2.18. The city of Watsonville suffered major liquefaction and settlement during the 1906 San Francisco earthquake. (Photo courtesy of the United States Geological Survey).

In the Pajaro Valley, particularly along the river, numerous fissures opened up. At the Granite Rock Company (Logan Quarry), right on the fault line, a rock crusher toppled onto a train and wrecked several cars. *"On the higher ground between Watsonville and Aptos, the shock was little felt. There was no movement along Aptos Creek, both wagon and railway bridges were unaffected".*

San Lorenzo Valley and the Santa Cruz Mountains

Damage in the valley and in the mountains seemed to primarily be related to the underlying foundation material, with some of the communities along the river bottom (Boulder Creek) suffering far more than areas like Bonny Doon, underlain by bedrock. The ridge tops, however, which also lie along the fault trace, did experience severe shaking.

"In the town of Boulder Creek, all chimneys were down except those on some 1 story cottages; these were cracked, however. People generally ran out of doors, but were not as a rule very badly frightened; some even stayed inside until they had dressed.

"At Ben Lomond no fissures nor other such evidences of the earthquake were to be seen. Inquiry showed this condition to continue in the country about the town. Broken chimneys were the only evidence.

"In this village (Felton) *the shock was apparently lighter than at either Boulder Creek or Ben Lomond.*

"On the road thru Bonnie (sic) *Doon the shock was uniformly light; chimneys were unharmed, plaster was intact, clocks did not stop, and even the milk had not spilt* (sic) *from the pans. People did not run outdoors. A top-heavy and rickety pigeon-house did not fall over, tho shaken considerably.*

"At the Wilder Dairy, on the Santa Cruz Pescadero road, 2 miles west of Santa Cruz near Meder Creek, the damage done by the shock was in the form of broken chimneys and cracked plaster in the houses."

Summit Ridge

"At Summit, a summer resort, the new hotel and several small cottages were all thrown toward the north. The main fault fracture is about 500 feet northeast of the hotel, and a secondary crack close to it has a downthrow of from 5 to 7 feet on the north or downhill side.... The Summit schoolhouse was dropt (sic) *4 feet downhill from its original position...All brick chimneys on the ridge fell, mostly to*

the north...The banks of Burrell Creek appear to have approached each other, so that the creek has become very much narrower (This was no doubt due to landslides or hillside failures, which moved material downslope into the stream bottoms.). *Water pipes were broken and twisted and filled with dirt... The Morrell ranch is located 1 mile south of Wright's Station and is on the line of the fault. The house itself was exactly upon a fissure, which opened up under the house at the time of the earthquake (See Figure 2.8). The house was completely wrecked, being torn in two pieces and thrown from its foundation...At Freely's place, 4 or 5 miles north of Morrell's, some 15 acres of woodland have slid into Los Gatos Creek, making a large pond. There are many other slides in the neighborhood and many broken trees.*

"The ridge on which we camped was full of cracks, ranging up to 2 and 3 feet in width, and in length from a few rods (40-50 feet) *to 0.25 miles; all trending west of north to northwest. All chimneys on the ridge were thrown down; several houses were completely wrecked; branches were broken from the trees, while many of the trees broke in two and others were uprooted. The canyon south of us was filled with landslides...I obtained a small bottle of crude oil from Mr. Sutton, which he said was dipt* (sic) *up from the ground on his neighbor's ranch, several hundred gallons of oil having run out of the ground since the earthquake, where there had been no sign of oil before".*

Perhaps the most tragic and devastating events in the county were the landslides and mud and debris flows resulting from the earthquake. Although there are many similarities in the effects of the 1906 and 1989 earthquakes on the lowlands, the effects in the mountains were very different due to the differing rainfall patterns. The April 1906 quake followed a period of above average rainfall, whereas the 1989 shock came after 3 years of drought. The excess water in the soils at the time of the 1906 quake led to large landslides and debris flows which moved considerable distances downslope.

On Hinckley Creek, a tributary of Soquel Creek, a landslide 500 feet wide, extended all the way to the ridge top, descended with *"extraordinary speed"* burying the Loma Prieta lumber mill *(Figure 2.19)* under a mass of rock and trees. *"The mill, boarding house, and other buildings of the plant were situated in a gulch, and were overwhelmed by a portion of the mountain – 1500 feet long, 500 feet wide and 100 feet deep, which slid down on top of them. The mill and everything in the gulch were forced up the opposite slope of the mountain and there buried to a depth of one hundred feet."*

FIGURE 2.19. The Loma Prieta Mill on Soquel Creek before being destroyed by the Hinckley Slide during the 1906 earthquake. (Photo courtesy of the Advocates for the Forest of Nisene Marks).

Nine men were buried instantly, while others, only several hundred feet away, were spared. The landslide dammed the stream, forming a lake up to 100 feet deep. Hundreds were involved in a massive digging effort in the following week, but only three bodies had been discovered after five days of searching. More than a year passed before the last body was recovered. The dead were listed in the San Francisco Examiner of April 24, 1906 as: *"J.J. Walker, Fred Peasle, J.O. Dunham, A. Buckley, H.W. Estrada, Frank Jones, Alexander Morrison, August Vollant and a Chinese."*

"On Deer Creek (a tributary of Bear Creek in the San Lorenzo Valley) *a large landslide started from near Grizzly Rock and slid westward, but changed its direction 60° or more farther toward the creek (Figure 2.20). The mill in the creek bottom below the slide was partly buried, and one man was killed. It is 500 feet from the mill in the gulch to the top, at the point where the slide started. The slide covered about 25 acres of ground, and destroyed a lot of virgin timber from 3 to 10 feet in diameter. The slide material, which is 300 feet deep, is composed of soil, clay and shale."*

FIGURE 2.20. The Deer Creek Landslide in the San Lorenzo Valley during the 1906 earthquake partially buried a lumber mill and killed one man. (Photo courtesy of the University of California Santa Cruz Special Collections).

An additional account in the San Francisco Examiner reported that James Dollar and Frank Franklin of Boulder Creek were buried alive in the Deer Creek landslide at Harman's mill. Additionally, *"On Bear Creek a smaller slide had moved a few hundred feet, buried a hut and killed one man".*

MONTEREY COUNTY

Monterey, Pacific Grove and Carmel are all built almost entirely on the granite of the Monterey Peninsula and damage was very minor during the 1906 earthquake on the peninsula. They are also farther from the San Andreas Fault and the 1906 rupture zone. Only several chimneys were damaged in Pacific Grove although shaking was moderate to severe according to residents. The Pacific Grove lighthouse was built on sand dunes and suffered greater damage as a result, with cracking of the dome structure. Monterey experienced essentially the same intensity of shaking as Pacific Grove, with no apparent damage done to houses and the only losses reported were some broken glassware in stores and some top heavy furniture overturned.

The Del Monte area suffered to a greater degree, particularly the Del Monte Hotel, which was built on alluvium and fill and was surrounded by marshy land, ponds and sand dunes. *"There were over 50 chimneys in the hotel, and half of them were thrown down, one crashing thru the roof on the west side of the hotel and causing two fatalities (Figure 2.21). The chimneys were tall and top-heavy, having ornamental tops; and while the damage to the interior of the hotel was very slight, showing that the earthquake was not of a violent type, the vibrations were sufficient to throw these top-heavy chimneys.*

"On the road eastward to Salinas from Del Monte, no visible signs of the earthquake were encountered until the Salinas River was reached. The Salinas Bridge was moved southerly several feet....so as to render the bridge unsafe".

In the fields Between Monterey and Castroville, geysers were reported that extruded *"boiling hot, bluish...mud to a height of ten to twelve feet. In places these geysers are from four to ten apart and in other sections they are fifty feet or more apart. The railroad tracks for almost the entire distance are twisted and* (lost) *all semblance of tracks... Near Castroville, while the disturbance was at its height, Foreman H.J. Hall grabbed his two children...and as they passed through the door they saw the earth open up and a crevasse, which Hall described as fully six feet wide, open and close several times....*" An observer that evening noted *"the house standing in a pool of geyser mud... like quicksand, and of unknown depth"*. All of these descriptions speak to the widespread process of liquefaction during the 1906 event along the lowlands next to the Salinas and Pajaro rivers.

FIGURE 2.21. The Del Monte Hotel in Monterey lost about half of its 50 masonry chimneys during the 1906 earthquake. (Photo courtesy of the Dudley Knox Library, Naval Postgraduate School).

Buildings and facilities out on the Moss Landing sand spit were extensively damaged due to the unconsolidated nature of the underlying sand and a high water table *(Figures 2.22 and 2.23)*. *"The wharf at Moss Landing bucked up and partly collapsed, while the warehouses were wracked and fell westward... the condition of the wharf indicates an eastward movement of the sand-spit. It is reported that at places along the pier where the water was formerly 6 feet deep, it now has a depth of 18 or 20 feet* (This no doubt was a result of slumping of sediments into the head of Monterey Submarine Canyon as also happened in 1989)...*Where there were sand dunes a few days ago, now there are deep holes with water bubbling through them...*

"North of Moss Landing the ground settled nearly 2 feet in places... as shown by the sagging of the (railroad) *track below grade line... The stretch of narrow-gage track parallel to the coast has been disturbed for nearly its entire length...the rails... were twisted into all conceivable shapes where not broken altogether.*

"At the hotel and stores on the mainland (at Moss Landing), *brick chimneys fell, but plastering was not seriously cracked...At Moss Landing, where the* (Salinas) *river runs parallel with the shoreline, the strip of land is seamed for miles. A crack, or rather a sink, about 20 feet wide and 4 or 5 feet deep ran under the buildings and rent them asunder. The office building between this crack and the river has been moved bodily – land and all – about 12 feet toward the river. Some of the cracks run into the ocean."* Until 1910 the lower course of the Salinas River flowed north, behind the sand spit through present day

Moss Landing Harbor, past the entrance to Elkhorn Slough, and discharged into the ocean north of the present Moss Landing Harbor entrance.

"The marshlands, river banks and some farm land along McClusky Slough and along the Pajaro River near its juncture with the Salinas River north of Moss Landing were extensively cracked.

"Fresh water came out of some fissures (Figure 2.24). The bottom of the Pajaro River came up to a point just north of its juncture with the Salinas River causing Pajaro River to change its course and empty into Monterey Bay near the present mouth of the river."

Despite being a hundred miles from the earthquake epicenter, the towns along the Salinas River valley suffered from severe seismic shaking, as did those valley floor communities to the north, such as San Jose and Palo Alto, which were also built on deep alluvium. *"The town of Salinas suffered greater damage than any other place in the county. Nearly every house and building was damaged to some extent. Plaster fell, windows broke, chimneys fell or were cracked, and brick buildings had their upper portions thrown off and, in some cases, almost completely demolished. The town is on the flat valley land, about 3 miles east of the river... the flood plain of the Salinas River was caused to lurch toward the stream from both sides.. these have the effect of landslide scarps and terraces (Figures 2.25 and 2.26)...numerous craterlets were formed by the sudden ejection of water from the un-*

FIGURE 2.22. Severe shaking and liquefaction of the soft sediments in the Moss Landing area led to failure of the ground and buildings during the 1906 earthquake. (Photo courtesy of the United States Geological Survey).

PHOTO: 2.23. Liquefaction and lateral spreading in the Moss Landing area in 1906 left this road cracked and useless. (Photo courtesy of the United States Geological Survey).

derlying sands (liquefaction) *due to the compressive action of the shock."* Near Spreckels *"water gushed forth at numerous places... and spurted repeatedly as high as 20 feet...for 10 minutes after the shock. The places...are marked by area of fine, light, bluish-gray sand, which is said to be known only at a depth of 80 feet in.. well borings of the vicinity". In these areas... are funnel-shaped depressions...from which the water issued."*

FIGURE 2.24. Liquefaction in the subsurface along the Pajaro River in 1906 produced sand volcanoes at the surface where the water saturated sediments were forced upward. (Photo courtesy of the United States Geological Survey).

The Spreckels sugar mill, was a very large (500 feet long, 150 feet wide and five stories high) steel and brick building on the south side of the river. The whole structure was shortened along its long axis, walls buckled and bulged, the ground outside heaved and deformed and water gushed from the ground. Damage was extensive *(Figure 2.27)*. Deformation, settlement and lurching of river bottom sediments continued all the way down the Salinas Valley to King City where the riverbed sank nearly 6 feet.

The disruption and damage caused by the 1906 earthquake were heavy in Santa Cruz County, although the only deaths reported were men buried beneath the large landslides in the Santa Cruz Mountains. Damage along the coastal area of Monterey County was much less severe, although the Salinas Valley communities did get hit harder. The event sparked considerable interest in earthquake research and data collection that changed our perspective of the role and importance of the San Andreas Fault in California's geological evolution. Although considerable information was collected and many observations documented, our limited knowledge of such phenomena at the time of the quake, as well as the logistical difficulties of getting through the region and the lack of trained geologists, probably allowed much valuable information to go unrecorded. An example of some of the thinking of 112 years ago is demonstrated in the following quote from the Santa Cruz Sentinel by a so-called *"expert"* of the time, Dr. J.F. Frisbee: *"Earthquakes form on the border of sea and land. Earthquakes of an explosive variety* [referring to the April 18, 1906 event] *are caused by the production of steam, deep down in the earth".*

MARCH 10, 1910

An earthquake of estimated magnitude 5.8 that was felt over 50,000 square miles, was believed centered near Watsonville (or under Monterey Bay?) and described as a "*slow rocking motion of alarming force*". Most of the damage was centered in Chittenden where houses cracked and bottles were thrown off shelves. In Santa Cruz, articles fell off shelves and plaster cracked in the lower parts of town (the Pacific Garden Mall area on the San Lorenzo River flood plain).

FIGURE 2.25. Scarps from settlement along the Salinas River from the 1906 San Francisco earthquake. (Photo courtesy of the United States Geological Survey).

FIGURE 2.26. Settlement and lurch cracking from 1906 earthquake along the Salinas River. (Photo courtesy of the United States Geological Survey).

OCTOBER 22, 1926

Two earthquakes (with estimated Richter magnitudes of 6.1), centered in Monterey Bay, shook the central coast in October 1926, and were felt from Cloverdale in Sonoma County to Lompoc in Santa Barbara County. Damage in Santa Cruz consisted of toppled chimneys, cracked plaster, broken windows on Pacific Avenue, and structural weakening of brick buildings. To the north, the city water main was broken at Laguna Creek, and at Davenport, groceries were thrown from the shelves.

These moderately large earthquakes took place on a branch of the San Andreas system, now recognized as the San Gregorio-Hosgri Fault. This rift splits off the San Andreas north of San Francisco near Bolinas and has been mapped along the coastline all the way to Diablo Canyon in San Luis Obispo County. In 1970 we had no offshore information and this fault was recognized on land only as two short segments, one cutting across the Año Nuevo peninsula (the San Gregorio Fault) and a segment to the north passing along the edge of the Half Moon Bay airport, the Seal Cove Fault. With the proposal to build what would have been the nation's largest nuclear power plant at Davenport in 1969, geologic investigations were initiated in order to determine whether there were any earthquake risks along this stretch of California coast. Offshore seismic surveys were carried out, which showed this newly recognized fault extended for over 300 miles along California's central coast, connecting the San Gregorio and Seal Cove Faults and passing across Monterey Bay *(Figure 2.28)*. While less active than the San Andreas Fault, the 1926 tremors and a number of smaller shocks indicate that the San Gregorio-Hosgri Fault is active and, based on its total length, capable of generating earthquakes up to perhaps 7.5 magnitude.

APRIL 23-25, 1954

Two moderate earthquakes occurred over 3 days time in the spring of 1954, which were centered in the Gilroy/Hollister areas. The second was slightly stronger and did considerably more damage. East of Watsonville, several homes were seriously damaged, the ground cracked, chimneys fell and windows broke. Considerable damage occurred in the city of Watsonville itself where a water main broke, ground cracks opened up, a flagpole toppled, plaster cracked and the concrete fell off the face of a building. Merchandise fell off shelves in Capitola and Aptos

APRIL 24, 1984: MORGAN HILL EARTHQUAKE

A magnitude 6 earthquake centered in the Morgan Hill area at 1:15 p.m. on April 24, 1984, was felt up to 400 miles away, shaking tall buildings in San Francisco and as far away as Sacramento and Reno. The quake and its dozens of aftershocks were actually on the Calaveras Fault. Shaking in some areas was reported as lasting for nearly 20 seconds. Most of the damage was reported in Morgan Hill where four houses were shaken off their foundations, and 23 others were damaged in the hillside subdivision of Jackson Oaks. In downtown Morgan Hill, four mobile homes were knocked off their foundations, and many storefront plate glass windows were broken as contents fell off supermarket shelves. An old adage was proven once more that during an earthquake in a supermarket, boxes of Wheaties survive better than bottles of Zinfandel.

FIGURE 2.27. Partial collapse of Spreckels Sugar Mill near Salinas during 1906 earthquake. (Photo courtesy of the University of California Berkeley, Bancroft Library).

Shaking was strong enough in Santa Cruz to empty the shops up and down the Pacific Garden

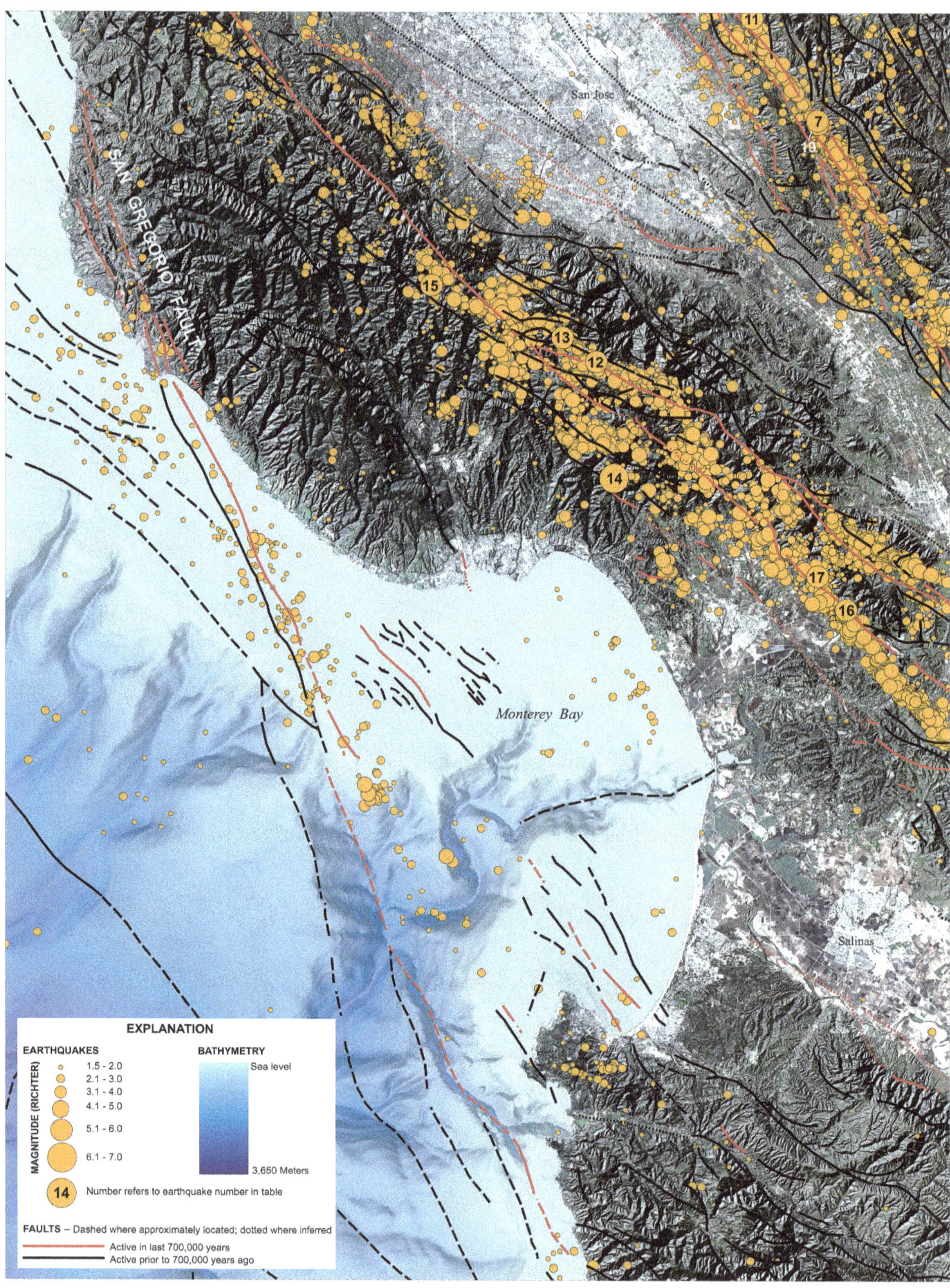

FIGURE 2.28. Active faults and recent earthquakes between the San Francisco Bay and Monterey Bay from 1970 – 2003. (Illustration courtesy of the United States Geological Survey).

Mall. A plate glass window in the old Woolworth's lunch counter exploded into the street, as shelves in supermarkets, liquor and drug stores collapsed, spilling contents onto the floors. Plaster fell from the walls of several downtown buildings and at least one water pipe broke, flooding a first floor.

FIGURE 2.29. Liquefaction under the Moss Landing Marine Laboratories during the 1989 Loma Prieta earthquake led to the complete loss of the facility.

OCTOBER 17, 1989: LOMA PRIETA EARTHQUAKE

While there is no one alive today who experienced the 1906 San Francisco earthquake, many of us in the Monterey Bay region have vivid memories of the October 17, 1989 shock. We can remember exactly where we were, whom we were with, and what we were doing. At 5:04 pm on that warm October evening, 15 seconds of shaking forever changed the Monterey Bay region. Three people died in the collapse of older downtown brick buildings in Santa Cruz as the walls and ceilings collapsed. Survivors dug vainly with bare hands at the shattered timber and bricks in futile rescue attempts at the Santa Cruz Coffee Roasting Company. A grandmother died when the brick walls of the Bake Rite Bakery in Watsonville collapsed, and two others were killed in unlikely situations; one man died in his pickup truck on Highway 1 as a spooked horse running along the freeway collided with his vehicle, and another was buried by a rock fall on a north coast beach. The sand spit under the Moss Landing Marine Laboratories liquefied, as it did in 1906, splitting the buildings down the middle and completely destroying the facilities *(Figure 2.29)*. The Monterey Peninsula and Monterey County, in general, suffered relatively little damage.

As evening fell, many throughout Santa Cruz County camped outside in their yards and in vacant lots as a seemingly endless series of hundreds of aftershocks continued to shake the region and its frightened residents. Initial reports indicated that most of downtown Santa Cruz had suffered major damage and would have to be torn down. Inspections followed inspections and in the end, many of the buildings did ultimately come down. The costs of repair and seismic upgrading were determined to be greater than building new safer structures.

The 6.9 magnitude 1989 earthquake was one more in a series of disasters over the past 175 or so years that have included earthquakes, floods, and fires, and which have wreaked havoc on downtown Santa Cruz. After an earthquake, the choice for reconstruction material was typically wood. After the next fire, masonry was favored. While Cooper House was damaged in 1906, it was upgraded and remodeled in the 1970s and become a popular Pacific Garden Mall attraction. While it did not collapse in October 1989, the earthquake damage and cost of restoration was deemed too high and it sadly fell to the wrecking ball a week after the quake. In the first month, 20

FIGURE 2.30. Many of the older unreinforced masonry buildings in downtown Santa Cruz were heavily damaged during the Loma Prieta earthquake.

FIGURE 2.31. Chimneys on many of the older houses in downtown Santa Cruz had no steel reinforcing rods and failed in the 1989 earthquake.

FIGURE 2.32. A number of houses in downtown Santa Cruz had "soft" lower sections with inadequate shear bracing that led to partial failure.

downtown buildings were demolished and more were to follow *(Figure 2.30)*.

Nearly every house in Santa Cruz suffered some damage. Many older chimneys either fell *(Figure 2.31)* or were cracked at the roofline and were ultimately rebuilt. Had October 17, 1989 been a cold fall day with wood burning in fireplaces and woodstoves, fires would no doubt have started in many homes as chimneys collapsed and caused far more damage. It might have become the Loma Prieta Earthquake and Fire, as the 1906 San Francisco event became so as not to discourage rebuilding and visitors. Fifty homes were destroyed in the city and many other older homes were damaged, with the downtown flood plain area, particularly Myrtle, Laurel, and Blackburn streets being hardest hit *(Figure 2.32)*. Estimates of home damage in the city reached $138 million (in 2017 dollars). Downtown business came to a halt as safety fences were placed around the hazardous and condemned structures, which at one point included 32 square blocks. Within a few weeks, a decision was made to construct a number of large tents or pavilions in the city parking lots to house many of the downtown businesses until repairs, retrofitting or reconstruction could take place. The subsequent months were challenging for many business owners. At the time, the Loma Prieta earthquake was one of the costliest natural disasters in U.S. history. Even with an epicenter in a remote area, in the heavily forested southern Santa Cruz Mountains, it still caused major damage and loss of life over 60 miles away in San Francisco and Oakland. In all, sixty-seven people died, 3,757 people were injured, 18,306 homes and 2,575 businesses were damaged and total damage was over $11 billion (in 2017 dollars).

The Loma Prieta earthquake ruptured a 25-mile long segment of the San Andreas Fault beneath the central Santa Cruz Mountains southeast of Highway 17. This segment had been recognized as having the greatest chance for producing a magnitude 6.5 to 7 earthquake of any fault segment in California north of the Mojave Desert. While the location and magnitude of this earthquake were not surprising to geologists and seismologists, there were no obvious short-term precursors to warn of the impending earthquake. The Pacific Plate (or Santa Cruz side of the fault), moved northwest about 6.2 feet and upward about 4.3 feet, relative to the North American Plate on the other side (San Jose side of the fault).

An unusual aspect of the earthquake was the absence of recognizable surface rupture along the fault itself. From observing the effects of earthquakes of this magnitude elsewhere in the world, geologists expected that fault rupture at the surface would have occurred; instead, a 3-mile wide zone of cracking was observed along the general trend of the fault along the crest of the Santa Cruz Mountains *(See Figure 2.9)*. Offset or separation along many of these cracks was great enough to damage houses, roads, and utilities. This type of widespread ridge top cracking represented a previously unappreciated earthquake hazard, one that extended well beyond the usually well-defined fault trace itself. Homes with the most spectacular views, those built on narrow ridge tops, consistently suffered the greatest damage from ridge top cracking, particularly along the Summit area southeast of Highway 17. Pinecrest Road, Blue Ridge Road, Rebecca Drive, Laurel and Comstock Roads, Stetson Road, Amaya and Longridge Roads were all sites of major structural damage due to severe shaking and ridge top cracking *(Figure 2.33)*. Many newer houses suffered complete failure due to a combination of lack of adequate connection to foundations *(See Figure 2.14)* or insufficient 1st floor shear bracing on two-story houses, particularly those with large two-car garage openings *(Figures 2.34 and 2.35)*.

Many of these cracks in the Summit Ridge area southeast of Highway 17 were concluded to be the result of reactivation of large ancient land-

FIGURE 2.33. The ridge tops in the Santa Cruz Mountains underwent severe shaking due to their proximity to the San Andreas Fault zone during the 1989 earthquake. Many houses built on ridge tops, particularly those without adequate seismic or lateral bracing or tie-downs, suffered significant damage or were complete losses.

FIGURE 2.34. Collapse of the 1st floor of a ridgetop home in the Santa Cruz Mountains during the Loma Prieta earthquake due to inadequate bracing.

FIGURE 2.35. A ridgetop home where failure of an inadequately braced 1st floor garage wall led to collapse of overlying structure in 1989 earthquake.

slides, similar to observations recorded in 1906. An important difference, however, was the very dry conditions in October 1989, in contrast to the saturated ground in April 1906. The hillsides in 1906 turned to mud and flowed downhill for hundreds of feet, carrying tall redwoods and destroying or burying everything in their paths. In 1989, the 15 seconds of strong shaking dislodged rocks and loose material, which slid or rolled downslope, and also disrupted the old landslides, causing cracks around their edges, but produced little overall movement.

As predicted in various hazard maps, a repeat of the 1906 earthquake damage took place where seismic shaking was amplified in areas underlain by thick deposits of water-saturated, uncompacted sand and mud. Downtown Santa Cruz and Watsonville, San Francisco's Marina district and similar settings around the margins of San Francisco Bay suffered from enhanced shaking, liquefaction, sinking and building tilting. This was precisely what had happened in these same areas in 1906. Many older building had never been brought up to modern seismic codes. The lack of reinforcing steel and the deterioration of the mortar between the bricks in older masonry buildings all contributed to the damage.

Several large fires broke out in Watsonville and thousands of the city's 30,000 residents were dislodged from their apartments and homes. Makeshift tents sprung up in vacant lots and in parks that first night while many others slept in their cars. The earthquake destroyed 195 homes and damaged over 1100 more, many in the poorest sections of town, where farm laborers, packing plant workers, the unemployed and the elderly resided. One hundred and six mobile homes were completely destroyed in mobile home parks as they fell of their jacks, while another 340 were damaged. The predominantly brick downtown area suffered as 32 business properties, many historic buildings, were seriously damaged. Nine hundred injured people were treated at Watsonville Community Hospital in the days immediately following the shock. The city's estimated damage to homes amounted to $302 million while businesses suffered an additional $358 million. Losses of frozen food and farm damages reached $138 million (all in 2017 dollars). With the collapse of State Highway 1 Bridge over Harkins Slough *(Figure 2.36)*, all of the freeway traffic was directed through downtown Watsonville, which added to the problems.

FIGURE 2.36. Collapse of State Highway One Bridge over Harkins Slough near Watsonville during Loma Prieta earthquake due to severe seismic shaking. Note that concrete caissons have pierced the asphalt on the bridge. (Photo: Jeff Marshall © 1989).

Although brief references to coastal bluff failure appeared in accounts of earlier earthquakes, there were very few homes that had been built close to the coastal bluffs at the time of the last large earthquake in 1906. Residents a century ago seemed to place less value on building as close as possible to the bluff edge than do today's property owners.

Cliff failure took place during the Loma Prieta earthquake along about 100 miles of coastline,

FIGURE 2.37. This Place de Mer townhouse development was built at the base of an ancient sand dune. During the 1989 Loma Prieta earthquake major sections of the bluff failed leading to partial failure of the foundation and then demolition of the house at the crest of the bluff and also partial blockage of the access road to the homes at the base of the bluff.

from Marin to Monterey counties. Bluff collapse and damage were greatest between Santa Cruz and La Selva Beach, however, closest to the epicenter. Shaking produced cracking of the bluff edge, which damaged bluff top homes *(Figure 2.37)*. As the loose debris cascaded downslope it damaged and blocked access to homes at beach level in places like Las Olas Drive, Beach Drive, and Place de Mer. Three blufftop homes in the Rio del Mar area and six Capitola apartments on Depot Hill were subsequently demolished as a result of earthquake damage and foundation cracking *(Figure 2.38)*.

There are not many events which are fixed in time as precisely as a big earthquake. The town clock in Santa Cruz stopped at 5:04 and stayed that way for months. The Loma Prieta (dark mountain or hill in Spanish, dark indeed) shock was the largest California earthquake in 37 years and the strongest on the San Andreas since the 1906 San Francisco event. Twenty-five miles of the mythical San Andreas Fault in Santa Cruz County had ruptured, from Highway 17 southeast to Pajaro Gap. The Santa Cruz side and the Pacific Plate moved about six feet to the northwest and rose vertically about four feet relative to the San Jose side in those 15 long seconds. The starting point for the shaking was about 11 miles beneath the Forest of Nisene Marks, five miles from the center of Aptos, which also suffered serious damage. For comparison, the 1906 earthquake ruptured 230 miles of the fault, from San Juan Bautista in the south to Pt. Reyes in the north. The Santa Cruz Mountains segment had been locked for 83 years, slowly accumulating strain, until 5:04 p.m., October 17, when it broke loose.

Epicenters are hypothetical points placed on maps to give us some sense of where earthquakes took place. Nearly all earthquakes occur at some depth within the Earth, and epicenters are defined as the location on the Earth's surface above the place where the initial failure or rupture took place, which is called the hypocenter or focus.

FIGURE 2.38. Cliff failure along Depot Hill in Capitola during the Loma Prieta earthquake led to partial loss of support for six units of the Crest Apartments and subsequent demolition.

There is considerable uncertainty in locating epicenters, however, and in some ways it's as much an art as a science. Determining the location of an epicenter depends upon analyzing the records of a number of seismographs at different locations. The job of a seismologist is to study the seismograph record and measure the gap or difference in the arrival times of different types of seismic waves, which are related to the type of rocks between the earthquake and the seismograph. We don't know the distribution of different types of rocks beneath the ground surface so there are some assumptions involved, which often produce significant uncertainties in precisely locating the epicenter. This is the best we can do, but it is imprecise nonetheless. This lack of certainty, however, doesn't discourage many curious people who have a strong desire, almost a magnetic attraction, to get to that place on the map with the X on it.

So, some enterprising individual(s) decided to mark the approximate epicenter of the Loma Prieta earthquake in the Forest of Nisene Marks, which is a well used, but somewhat out of the way, state park. The sign and location were well publicized, which brought literally thousands of people into a remote section of the park. Some visitors arrived in limos and fancy clothes, having decided to drop by on the way to a fancy dinner party. Park rangers, realizing that this attention and traffic was taking its toll on the trail to a rather remote area of the park, decided to relocate the epicenter sign to a more accessible location. The new epicenter was no doubt just as accurate as the original site, and visitors received just as much satisfaction in reaching the new *"epicenter,"* their seismic pilgrimage now complete.

Although many central California residents fancy themselves as amateur seismologists, and immediately compare their magnitude estimates after each moderate tremor in the region, most were unprepared emotionally for the aftershocks that followed the 1989 event. A magnitude 5.2 aftershock came 37 minutes after the main event, and 33 hours later, a 5.0 shock again struck terror in those who were just beginning to calm down. A dozen magnitude 4 or greater earthquakes shook the area in first 7 hours. In the first two days, sixty-seven 3.0 or larger aftershocks terrorized those still in the county *(Figure 2.39)*. It takes a while for two very large tectonic plates to come completely unstuck and large numbers of aftershocks can be expected to accompany every big earthquake.

Magnitude	Number	Effect
5	2	Damaging
4	20	Strong
3	65	Perceptible
2	384	Not felt
1	1,855	Not felt
<1	2,434	Not felt
Total	4,760	

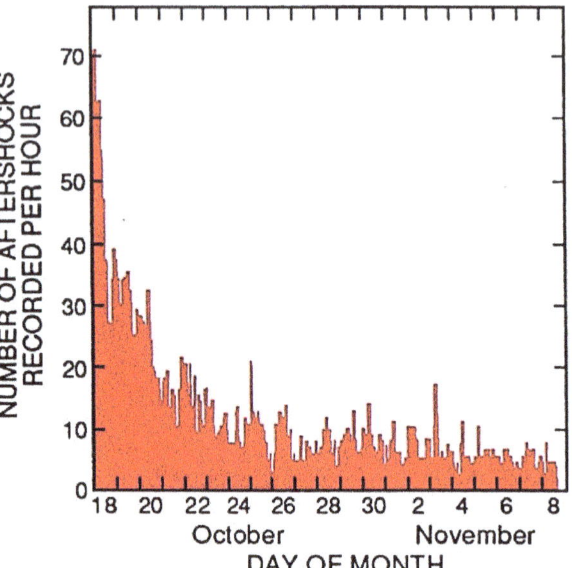

FIGURE 2.39. In the first three weeks following the October 17, 1989 earthquake, 4,760 aftershocks were recorded. (Table and graph courtesy of the United States Geological Survey).

SOME FINAL THOUGHTS ON EARTHQUAKES

The preceding historical account only describes the largest or most significant of the many earthquakes that have affected the Monterey Bay area over the past 190 years of written history. Dozens of smaller but perceptible shocks have also rattled the region, at least several every year. The patterns are not random. Earthquakes occur along relatively narrow zones, although they affect broad areas, and they occur when the rocks can no longer resist the accumulated strain they are accumulating. Based on our general understanding of plate tectonics and fault behavior, we believe that the 25-mile long San Andreas Fault segment that broke in 1989 should not produce another large earthquake for many decades. Geologists and seismologists may not all agree on this, but then they don't all agree on many things. However, the segment extending northwest from Highway 17 towards San Francisco has not broken since 1906. Similarly, the Hayward/Calaveras fault systems haven't ruptured with a large earthquake since the mid to late-1800s.

In their 2015 report, the Working Group on California Earthquake Probabilities estimated the likelihood or probability of earthquakes of magnitude 6.7 or greater, 7.5 or greater, or 8.0 or greater for the Northern San Andreas Fault, the Hayward and the Calaveras Faults in the San Francisco Bay area *(Figure 2.40)* in the 30 years from 2014 to 2044. Probabilities of a large earthquake on the San Andreas in the general area from Hollister to north of the Golden Gate are around one percent during this period, but higher on the Hayward and Calaveras faults, which are farther away. Based on these expectations and past measurements and observations, earthquakes of this magnitude on Bay area faults could produce significant shaking in Santa Cruz County, but not likely to be as severe as that which took place during 1989. Thus the worst may be over for us for a while and for as long as we don't move to Berkeley or somewhere else in the East Bay. Not that we won't have future earthquakes, but that shaking and damage in the immediate decades ahead shouldn't be any more serious that what we experienced in 1989.

Thus, there is some peace of mind, at least here in the Monterey Bay region. There is some other good news we may want to remember as well. Average annual worldwide death toll from earthquakes in the past century was about 20,000. In the United States, however, the average annual death toll was about 20, the great majority of these from the 1906 San Francisco earthquake (and fire). Information of this sort has led to the development of two general schools of seismic safety in the United States within which many professionals fall, the "*bathtub school*" and the "*what if I'm right school.*" The former consists of some renegade geologists and statisticians who state with statistics like those quoted above that you have a greater risk of dying in your bathtub than you do in an earthquake. The latter group consists of the civil defense specialists and many geologists and building officials who feel we should go to the effort to be prepared for the infrequent but inevitable future seismic events. And they both have good arguments. Without a doubt, the building codes now require far more seismic safety measures than were in effect 50 years ago when many older homes and other buildings were constructed. We have witnessed the types of failures that took place in San Fernando/Sylmar in 1971, Northridge in 1994, and Loma Prieta in 1989, and have addressed those by updating building codes for all new construction.

There may be some reduction in your seismic stress level to know that your chance of dying in

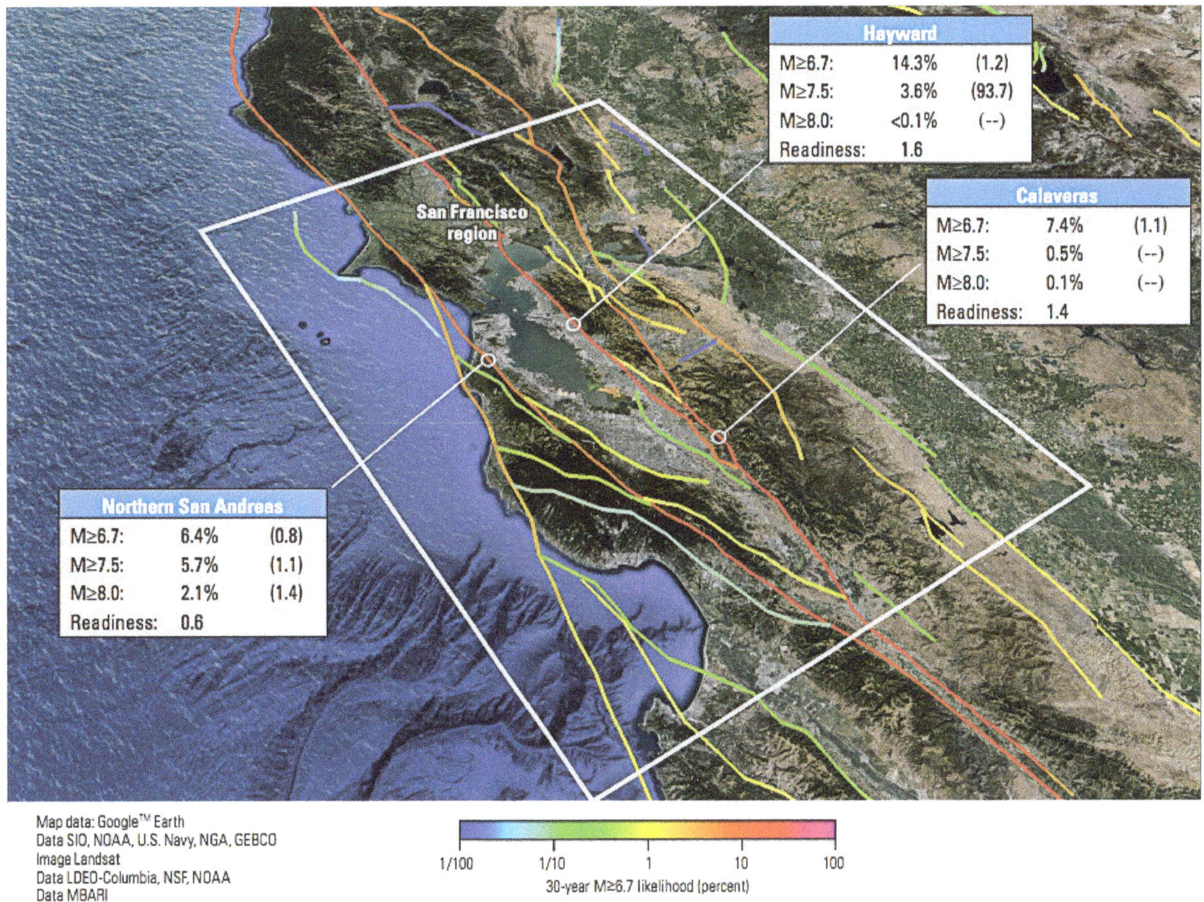

FIGURE 2.40. Central California faults and their probabilities of generating different magnitude earthquakes between 2014 and 2044. (Illustration courtesy of the United States Geological Survey).

many other ways is far higher. The odds of dying from an earthquake in this country are actually lower than those from lightning strikes, dog bites and insect stings. Odds of dying in a motor vehicle accident are about 1700 times greater, from falling 1400 times greater, and from drowning, 180 times greater. Yet many people spend far more time worrying about earthquakes then driving over Highway 17 or riding their bicycle and texting on a busy street. Cell phone distractions, whether talking or texting, cause a reported 2600 deaths and 330,000 injuries in the United States annually. Now that's a hazard we ought to be concerned about. There is no such thing as multi-tasking while driving an automobile.

During the Loma Prieta earthquake, one man died while sitting on the beach when a section of bluff collapsed. A second was killed when a runaway horse ran onto the freeway and collided with the front window of his truck. These were both bizarre accidents and were clearly low probability events. It should be very reassuring that no one, not a single person, died in that 6.9 magnitude earthquake in a single-family wood frame structure, although thousands were destroyed or damaged. In most of the rest of the seismic world, however, whether China, Pakistan, Iran, Turkey, Peru or Guatemala, most of the people still live and work in unreinforced or poorly reinforced adobe, brick, or masonry buildings which simply aren't able to stand up to severe seismic shaking. They don't have enough lumber to build with so they use what they have, and it hasn't worked very well.

While the odds of dying in an earthquake in this country are very small relative to almost any other risk, the property damage from earthquakes can be astronomical. The Loma Prieta earthquake produced about $11 billion in damage, but losses during the slightly smaller but more urban 6.7 magnitude Northridge earthquake of 1994 were nearly $20 billion, making it the most costly natural disaster in United States history at that time, and this wasn't even a large earthquake. For much of California, the worst may be yet to come. Hopefully for the Monterey Bay region, the worst is over, at least for the near future.

Chapter 3

Tsunamis

Watch for the tsunami. Everybody running 'cause the waves are coming...
– Frank DeLima

INTRODUCTION

Tsunami! This word often evokes an instant emotional response for most of us, much like earthquake, shark, or mountain lion. The massive 2011 earthquake and tsunami that produced death and destruction in Japan, and the 2004 earthquake and tsunami in the Indian Ocean that may have killed 235,000 people, were both tragic reminders of what can happen when waves 30 feet high wash quickly and often unexpectedly over densely populated coastlines.

Just as you can make little ripples by blowing on a cup of coffee, or create small waves by jumping into a swimming pool, any large disturbance or sudden movement in or under the ocean can also generate waves. Enormous oscillations of water caused by large seafloor earthquakes, underwater landslides, or exploding undersea volcanoes, result in waves known as seismic sea waves, or tsunamis. These waves have frequently been incorrectly called tidal waves, but they don't actually have anything to do with the tides.

The word *tsunami* is derived from two Japanese words: *tsu,* which means harbor, and *nami,* which means wave. Some of the greatest historical impacts of tsunamis have occurred in harbors or port cities where the wave energy has been concentrated as it reaches the shoreline. Hilo, Hawaii, is a good example. Situated at the upper end of a narrow bay on the big island of Hawaii in the middle of the Pacific, it's been repeatedly damaged by tsunamis from distant earthquakes around the margins of the Pacific Ocean *(Figure 3.1)*.

FIGURE 3.1. Waterfront area of Hilo, Hawaii, after the tsunami from Chile in 1960. (Photo Courtesy United States Navy).

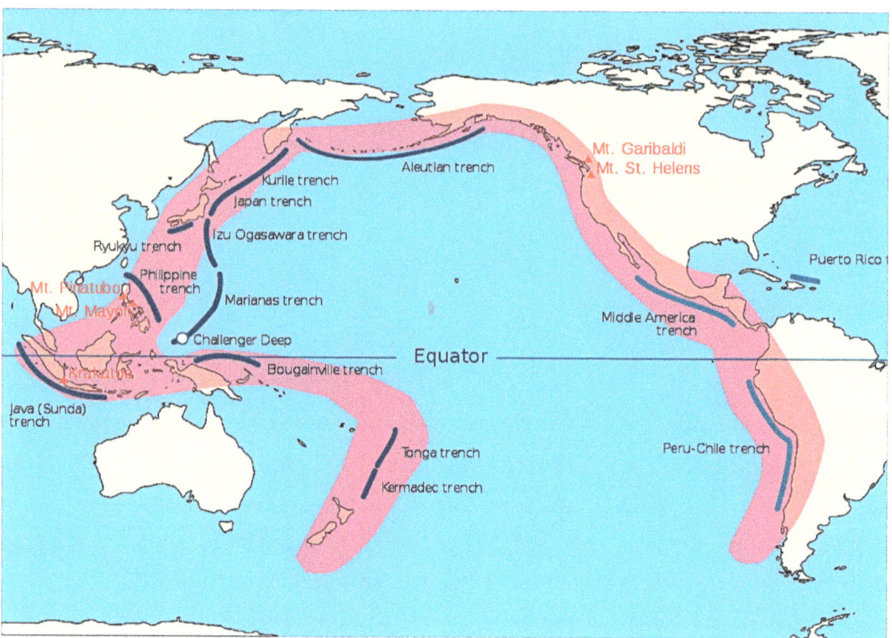

FIGURE 3.2a. Trenches around the Pacific Basin. (Image via Wikimedia - in the public domain).

SUBDUCTION ZONES AND TSUNAMIS

Many of the world's most damaging tsunamis have originated at trenches or *subduction zones*. The Earth's largest earthquakes occur at these locations where thin dense oceanic plates collide with and descend beneath thicker and lower density continental plates. All but a few of the planet's subduction zones occur

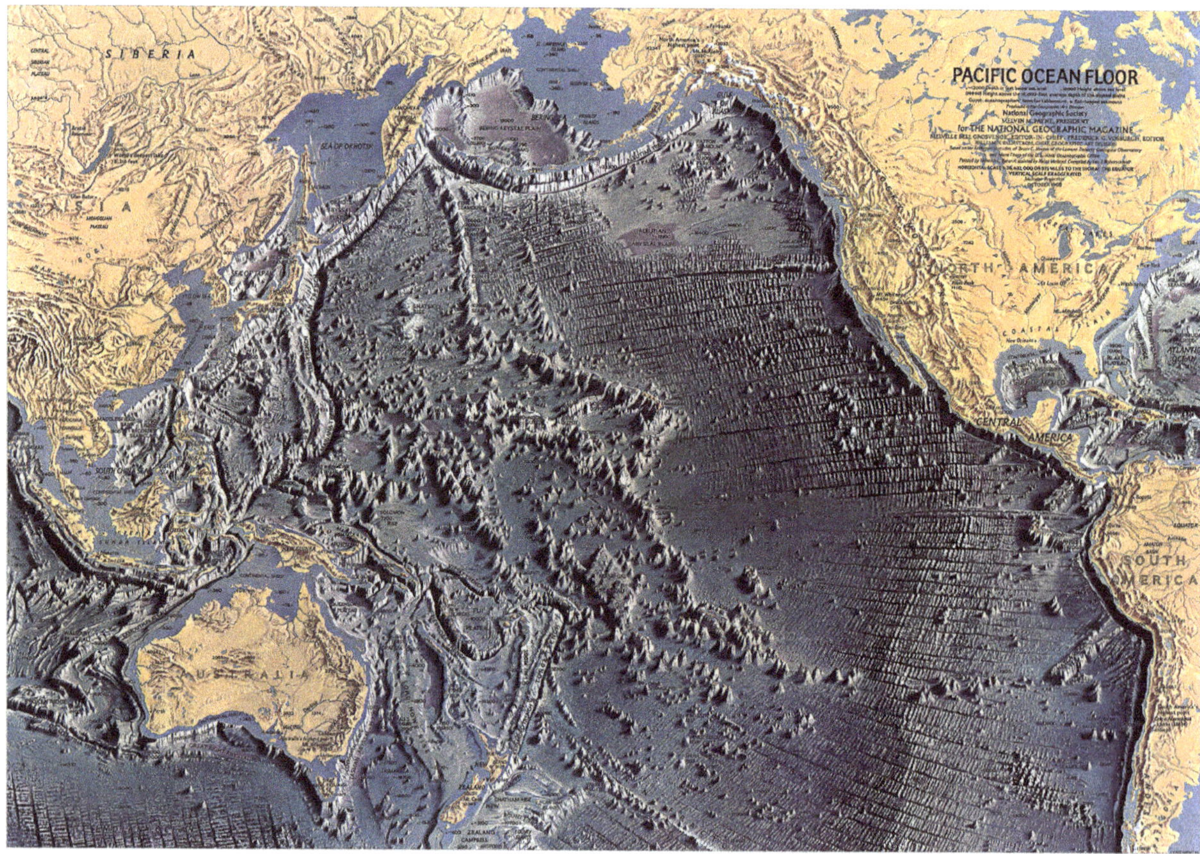

FIGURE 3.2b. Seafloor features of the Pacific Basin showing trenches nearly encircling the margins of the ocean basin. (Map courtesy of *National Geographic*).

around the margins of the Pacific Ocean – what has been called the "*Ring of Fire*" – and this is where most of the world's very large earthquakes and damaging tsunamis have originated. Moving counterclockwise around the Pacific Basin, we have the Peru-Chile, Middle America, Aleutian, Kurile, Japan, Isu Ogasawara, Ryuku, Marianas, Bougainville, Tonga and Kermadec trenches *(Figures 3.2a and 3.2b)*. And Hawaii, particularly Hilo, is right in the path of tsunamis generated at these trenches.

Everything about tsunamis can be described in superlatives. In the open ocean, these waves will typically travel at speeds of 450 or 500 miles per hour, about the same velocity as a commercial airliner. The wavelength of a tsunami, or the distance between two crests, is typically 90 to 100 miles, which contrasts to the typical wind waves that break along the central coast that range from several hundred to perhaps a thousand feet. With heights in the ocean of only several feet, however, and these very long wavelengths, tsunamis are essentially imperceptible and of no real concern to ships at sea. In fact, ships wouldn't even recognize or notice a tsunami while far from shore.

As these waves approach the coastline and move into shallow water, however, their speed and the distance between wave crests will diminish at the same time that their heights increase *(Figure 3.3)*. These long waves still have so much mass and momentum, however, that they will wash inland a significant distance and also rise to considerable elevations. During the great 2004 Indian Ocean earthquake and tsunami, waves washed inland for over two miles and at least one of the three waves reached an elevation of 100 feet above sea level. The 2011 tsunami in Japan was even more impressive as it flooded low areas up to six miles from the shoreline *(Figure 3.4)* and reached elevations as high as 135 feet above sea level.

Tsunamis can reach the Monterey Bay area from any of the trenches or subduction zones that nearly encircle the Pacific Ocean, from the Aleutians south to Chile, and from New Zealand north to Kamchatka *(See Figure 3.2a)*. So although these long-period waves might have traveled thousands of miles across the ocean from their source, because of the energy they carry, they can still move inland great distances, extend to considerable elevations above sea level, and cause massive damage and loss of life *(Figure 3.5)*.

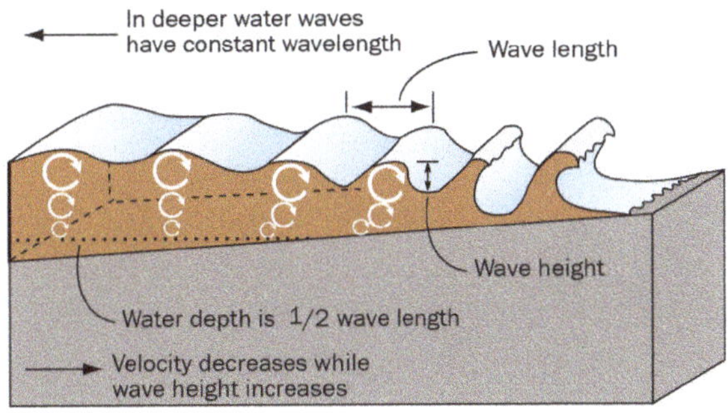

FIGURE 3.3. As waves (whether they are wind waves or tsunami waves) enter shallow water, wave lengths (the distance between two successive wave crests) are reduced, and wave heights (the vertical distance from wave trough to wave crest) increase. (Illustration courtesy of the University of California Press).

THE CASCADIA SUBDUCTION ZONE

The closest "*local*" source for a large tsunami is the Cascadia Subduction Zone that extends from offshore Cape Mendocino in northern California about 550 miles north to Vancouver Island *(Figure 3.6)*. At its closest point, the southern end of the Cascadia Subduction Zone and, therefore, the source of a large earthquake and tsunami, are about 300 miles north of Monterey Bay. Fortunately, however, because this subduction zone trends parallel to the coastline (as do almost all subduction zones), most of the energy from a tsunami generated will be transmitted at right angles to the zone, much like the beam of a flashlight. For Cascadia, most of the tsunami energy would move west, out into the North Pacific towards Japan, or east, directly on-

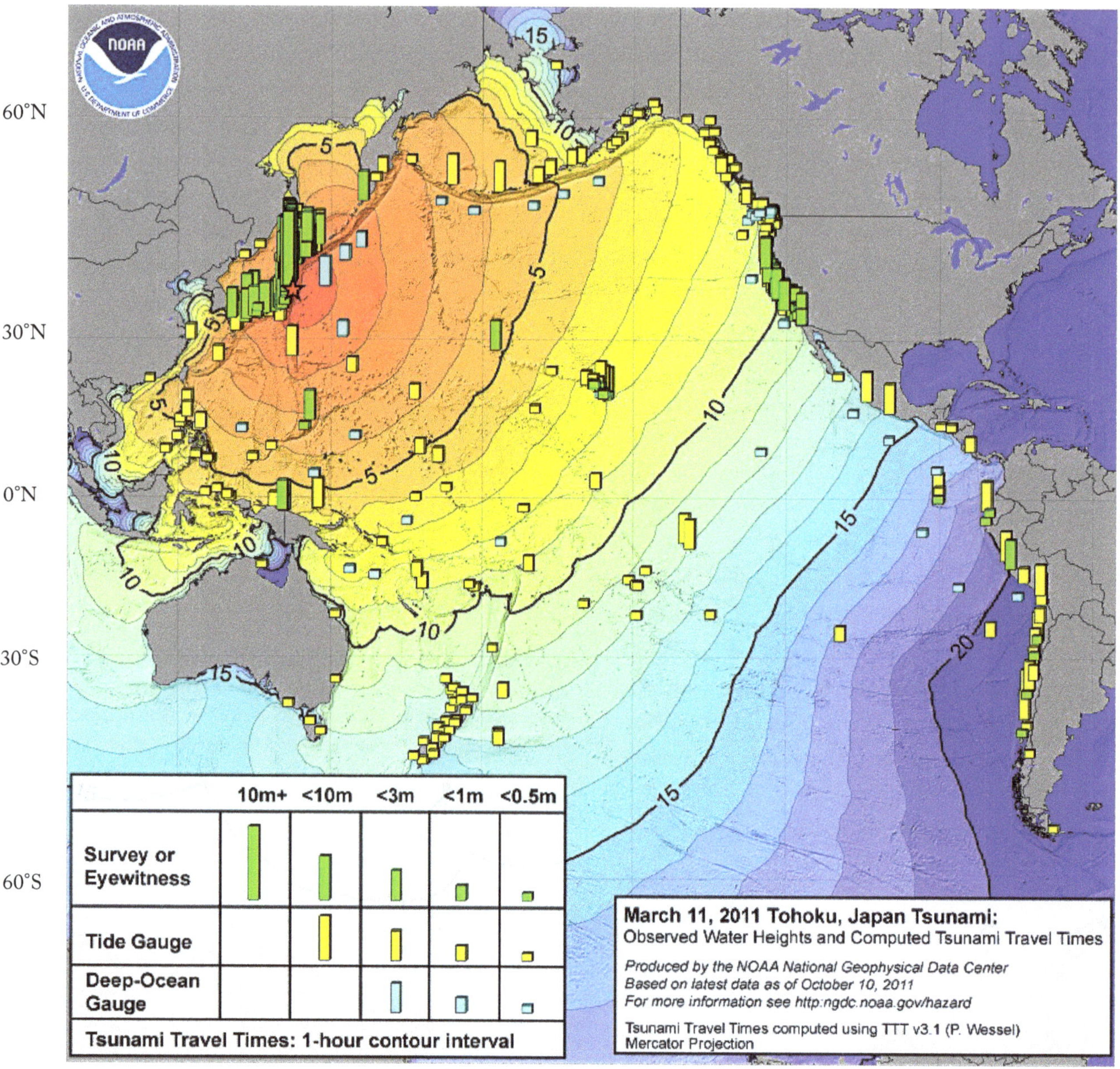

FIGURE 3.4. Heights of the 2011 Japan tsunami were greatest directly onshore, but were reduced in height as they propagated completely across the Pacific Ocean and increased as they washed onshore of the Americas. (Illustration courtesy of Katherine Mueller, IFRC, NOAA/NGDC).

shore, where damage and impacts will be greatest, and warning time shortest. Some of this wave energy will travel north and south (and in our case), down the California coast, its energy diminishing with distance.

Discoveries over the past 30 years by field geologists studying preserved sediments along the coastlines of northern California, Oregon and Washington have yielded evidence of a large tsunami that struck this area about 300 years ago. The offshore area is a boundary where one small plate, the Juan de Fuca, collides with the huge North American Plate and is forced down beneath the continent. As the Juan de Fuca Plate slowly descends, there is tremendous friction to a depth of several hundred miles as it scrapes against the bottom of the North American Plate and pulls it downward. Most of the time these two plates are locked, but when the accumulated stress is great enough for the two plates to shift and uncouple, the edge of overlying North American Plate will rebound and a very large amount of energy is released. The rebounding of a massive slab of seafloor when the plates detach displaces a large amount of ocean water, which typically produces a set of large waves or a tsunami.

There is mounting evidence along the coastline of the Pacific Northwest that very large sea floor earthquakes (magnitude 9) and resultant tsunamis occur every several hundred years in the offshore area between Cape Mendocino and Puget Sound. Some of the newer discoveries suggest that these huge waves moved a considerable distance inland into bays and estuaries and left behind clean beach sands within the

FIGURE 3.5. The March 2011 Tohuku, Japan, tsunami devastated more than 40 miles of coastline in northeastern Japan including Yamada Town. (Photo courtesy of Katherine Mueller, IFRC, NOAA/NGDC).

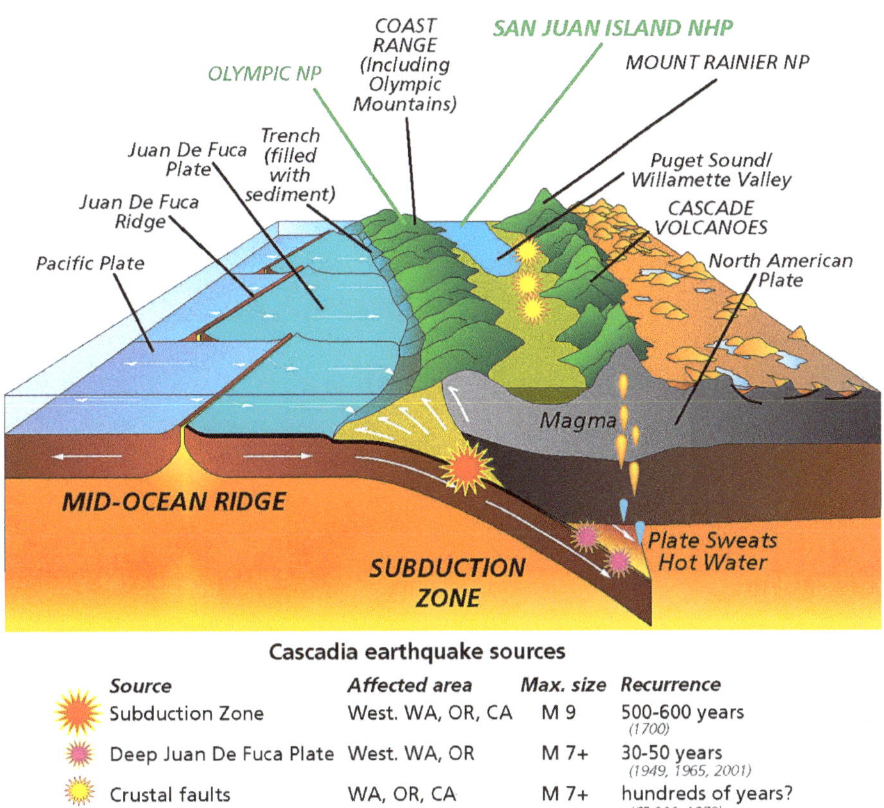

FIGURE 3.6. The Cascadia Subduction Zone extends from Northern California to Vancouver Island and has produced very large earthquakes every 300 to 500 years. (Illustration courtesy of R.J. Lillie and the National Park Service).

FIGURE 3.7. A ghost forest on the edge of an estuary along the Washington coast, where trees died as a result of salt water immersion from the January 1700 Cascadia Subduction Zone earthquake and tsunami. (Photo courtesy of Brian Atwater, United States Geological Survey).

muddy organic material that would normally be deposited in these more protected environments. In addition, during a major earthquake of this type, a large portion of sea floor and shoreline may suddenly move upward or downward. Large old cedar trees that formerly lived a few feet above sea level were submerged and died when their roots came into contact with salt water when the shoreline sank centuries ago *(Figure 3.7)*. They are still preserved along the coast of Washington. The growth rings on these old trees can be counted, and the age of the trees can also be documented using carbon-14 dating. From a number of investigations of this sort – studying the sediments left behind, the trees that died, and the ages of these materials – we have good evidence that large-magnitude earthquakes occur along the Cascadia Subduction Zone about every 300-500 years or so, on average.

Scientists have also recently uncovered written records at a monastery, 4500 miles away on the opposite side of the North Pacific, providing evidence that the last major earthquake on the Cascadia Subduction Zone generated a tsunami on January 26, 1700 that reached all the way to the coast of Japan. Confirming this date, and knowing that great Cascadia earthquakes and tsunamis occur about every 300-500 years, has significantly increased our level of awareness and concern about when another earthquake of this magnitude will occur, and how the associated tsunami might impact the Monterey Bay shoreline.

FIGURE 3.8. Tsunami warning signs are now widely posted along low-lying shoreline areas on the coasts of Washington, Oregon and California.

The California Office of Emergency Services, working with other state and federal agencies, has been developing tsunami risk or inundation maps for the populated areas of the central and northern California coastline, an effort that has also led to the posting of tsunami warning signs in most of the state's low-lying coastal communities *(Figure 3.8)*. The Cascadia Subduction Zone is less than 100 miles offshore of northern California, so for coastal cities like Eureka and Crescent City, any tsunami announcement would provide only minutes of warning time to coastal residents. We are fortunate here in the Monterey Bay region in having perhaps an hour or so of warning time between the occurrence of a very large earthquake and the arrival of a tsunami from Cascadia. Maximum elevations reached by historical tsunamis along the Monterey Bay shoreline from large earthquakes around the Pacific Basin over the past roughly 175 years of observations have been about ten feet on two different occasions (in 1946 and 1964, both from very large earthquakes in the Aleutian Trench off Alaska). The most extreme run-up conditions would occur when a tsunami arrives during a very high tide, which would lead to ocean water reaching higher elevations and extending farther inland.

For tsunamis generated from large earthquakes anywhere else around the Pacific Rim (the Aleutian Trench or Japan for example), our warning time is on the order of 5-8 hours. With a tsunami warning system now in place throughout the Pacific Basin, residents of the Monterey Bay area through their cell phones or other media, would be informed with ample time to make plans to evacuate very low-lying coastal areas.

POSSIBLE TSUNAMI FROM A LARGE LANDSLIDE OR SLUMP IN MONTEREY SUBMARINE CANYON

While some refer to the offshore Monterey Submarine Canyon as a trench, it is not a trench in the geological sense. It isn't a subduction zone where plates collide and large seafloor earthquakes occur; but it is a seafloor drainage system, where sand is transported from the shoreline at Moss Landing into very deep water miles offshore. Sea floor investigations using geophysical tools have revealed a number of large slump scars along some of the steeper canyon walls, indicating that very large failures occur from time to time. While there have been no direct observations of tsunamis initiated by such a process, and we have no idea when such an event may occur in the future, the possibility exists of such a local tsunami, most likely during a large earthquake on one of the active faults in the area. Models that have been developed to approximate a tsunami that could occur given the size of these ancient slumps indicate that a wave 3-4.5 feet high could reach the beaches around Monterey Bay in 6-12 minutes, not a lot of time but also not very large waves.

TSUNAMI HISTORY OF THE MONTEREY BAY REGION

Over the past nearly 200 years or so of somewhat reliable news reporting in California there have been 13-14 tsunamis recorded with heights of over three feet; just six of these have been considered destructive. Over this entire period, only 17 lives have been lost due to tsunamis, far less than the number of Californians killed by dog bites (six fatalities in 2016 alone) and almost any other accidental cause.

JUNE 15, 1896
JAPAN EARTHQUAKE AND TSUNAMI

What has been called the great Sanriku earthquake in Japan (magnitude 7.6) killed more than 26,000 people and produced a tsunami that arrived on the west coast on the United States. The *San Francisco Chronicle* of June 16 reported that a 9.5-foot wave arriving at low tide overtopped a temporary dike of sand bags protecting an area of the San Lorenzo River that was being used to build floats for the Venetian Water Carnival in Santa Cruz. The San Diego Union also reported that a five-foot wave destroyed a protective dike and that water rose far upriver and did severe damage to a ship moored at the pier. The Santa Cruz Sentinel had no report of any damage from this event.

APRIL 1, 1946
ALASKA EARTHQUAKE AND TSUNAMI

The two tsunamis that have been the most damaging to the California coast historically were both generated by very large subduction zone earthquakes in the Aleutian Trench off Alaska – they might be called the holiday earthquakes. The first took place on April Fools' Day in 1946 and was big, magnitude 8.6; and the other, an even larger 9.2 magnitude event occurred on Good Friday in 1964. The 1946 tsunami had modest impacts at Noyo Harbor at Fort Bragg in Mendocino County on the north, where many boats broke from their moorings, to Santa Catalina Island in the south. In Half Moon Bay, which sits at a very low elevation, 14-foot waves washed a quarter mile inland, damaging houses, boats and docks and also wrecking a fishing tackle shop in El Granada *(Figure 3.9)*.

Tidal Waves Wreck Seaside Fishing Tackle Shop

FIGURE 3.9. 1946 Alaskan tsunami damage at El Granada, near Half Moon Bay. (Photo courtesy of the UC Water Resources Center Archives).

Two waves, the first at 10:15 am and the second at 11:51 were reported in Santa Cruz with maximum heights later documented at about ten feet. The waves reportedly pushed water a considerable distance up the San Lorenzo River. On the municipal wharf, "*lines and buoys used to fasten fishing boats suddenly went taut and ladders down the pier into the water were lifted to a vertical position as the swell passed.*" Malio Stagnaro of the Stagnaro Fishing Corporation, reported that strong currents continued to agitate the area around the wharf the following day, preventing boat owners from anchoring their boats.

There was apparently only a single tsunami fatality along the entire coast of California from that large April Fools' Day earthquake and it just happened to be in Santa Cruz. It was also the only tsunami related death that the Monterey Bay area has ever experienced. For some historical perspective, in 1946 tsunamis were still called tidal waves and we had no plate tectonics to make sense of why these big earthquakes and tsunamis occurred. Additionally, we had no Pacific tsunami warning system as we do today.

About 10:15 am on April 1, 1946, Hugh Patrick, a 73-year old man walking along the shoreline near Lighthouse Point was drowned when the water level rose quickly to 10 feet above normal as the first wave hit. His walking companion, 73-year old Cephus Smith, described in the Santa Cruz Sentinel as "*a local dishwasher*", was also knocked over by the wave but was unsuccessful in an attempt to rescue his friend. Hugh Patrick's body was recovered in the kelp beds a half mile west of Lighthouse Point 17 days later. Another man, Ury Afanasief of San Francisco, was swimming when a surge dashed him against the rocks but he managed to fight his way out. Men on the municipal wharf reported the water receding at a terrific pace a little after 10:00 am and suddenly returning at an appalling speed and surging high on the beach. There were four surges, the last at 11:50 am, which all but topped the Esplanade along Main Beach. The bay presented a weird sight as it "*seethed, boiled and whirled.*" The *Watsonville Register-Pajaronian* reported that the wave covered the whole beach at Santa Cruz with ten feet of water. One observer reported that the sea rose ten feet above the normal level for the entire length of the wharf.

FIGURE 3.10. An announcement of an appearance of an adolescent evangelist in Santa Cruz, talking about the 1946 tsunami. (Image courtesy of the *Santa Cruz Sentinel*).

A week later the Santa Cruz Sentinel carried an announcement for an event featuring "*Little David*", an 11-year old evangelist ("*65 lbs. of fire*") along with Felicia Fernandez, a "*Mexican Prophetess*", speaking on "*Why the Tidal Wave in Santa Cruz and Will There Be Others?*" *(Figure 3.10).*

In Monterey some fishermen reported a slight turbulence in local waters but little else was noticed. In neighboring Pacific Grove a single surge of water was observed at the municipal swimming pool on the coast, which flooded the dressing room to a depth of three feet *(Figure 3.11)*. One man on the steps fled ahead of the water and photographic evidence indicates that the water reached an elevation of 10.3 feet above MLLW at Pacific Grove (MLLW or "*Mean lower low water,*" is the reference level for all tide tables, so this height would have been about 3.3 feet above the highest tides expected in Monterey).

NOVEMBER 4, 1952
KAMCHATKA EARTHQUAKE AND TSUNAMI

A magnitude 8.2 earthquake off the east coast of Kamchatka (the Kamchatka Trench and subduction zone) generated a tsunami that reached the entire west coast from Washington to California. One of the Stagnaro Company's fishing boats (the *Bruno Madre*) was damaged when a wave struck it while being hoisted onto the wharf. Waves rolled high onto Cowell's Beach, removing sand and leaving the lower end of Saunder's steps ten feet above the ground. Water rolled up to the scaffolding along the front of the Casino whose interior was being remodeled.

MAY 22, 1960
CHILE EARTHQUAKE AND TSUNAMI

A southern hemisphere great earthquake (8.6 magnitude) generated a tsunami that impacted the coasts of California, Oregon and Washington. The Santa Cruz Sentinel reported no damage along the Santa Cruz County coast from the

FIGURE 3.11. Municipal swimming pool at Pacific Grove where the high water from the 1946 Alaskan tsunami reached bottom of words BOATS and flooded the dressing room to a depth of three feet. One man fled up the steps ahead of the surge. (Photo courtesy of the Orville Magoon private collection).

"*tidal wave,*" but water washed up the steps of the Boardwalk Casino at 10:35 am and also crashed over the seawall at Capitola. Six-foot waves were reported by an observer on the Municipal Wharf arriving at 20-minute intervals through the morning. At Moss Landing, five-foot maximum waves were observed at 20-25 minute intervals along with severe currents in the harbor's entrance channel. The *Monterey Peninsula Herald* reported "*waves surging into the bay*" that completely submerged the partially completed seawall, but there was no visible damage. Water also rose to within a few feet of the city beach parking lot.

MARCH 27, 1964 GOOD FRIDAY ALASKA EARTHQUAKE AND TSUNAMI

Eighteen years after the destructive 1946 Alaskan earthquake and tsunami, the even larger 1964 9.2 magnitude Good Friday earthquake generated a tsunami that radiated out across the entire Pacific Basin from the Aleutian Trench and was the most destructive tsunami to batter California's coast in historic time. Crescent City, on the northern California coast, was the hardest hit, being inundated by a series of waves that pushed buildings off their foundations and into other structures, and swept vehicles and buildings into the ocean. *(Figure 3.12).*

FIGURE 3.12 Automobiles damaged during 1964 Alaskan tsunami in Crescent City, Humboldt County, California. (Photo courtesy of the Orville Magoon private collection).

Wave run-up extended 800 to 2,000 feet inland in the commercial and residential areas of the city with water depths of up to eight feet in city streets and 20 feet along the shoreline. The event and its impact were even memorialized on the wall of at least one commercial building *(Figure 3.13)*. The worst waves struck the waterfront area at 1:45 A.M., drowning 12 people, demolishing 150 stores, and littering the streets with huge redwood logs from a nearby sawmill. Most of the city's downtown was either damaged or totally destroyed, and rather than being rebuilt, the blocks nearest the harbor were subsequently made into a park (*Figure 3.14*).

FIGURE 3.13. A mural on the wall of a commercial building in Crescent City depicts the damage from the 1964 tsunami.

FIGURE 3.14. Following the extensive damage to Crescent City's business district from the 1964 tsunami the area was preserved as a park with no development. (Photo: Kenneth and Gabrielle Adelman © 2005).

According to the U.S. Army Corps of Engineers, property losses approached $218 million (in 2017 dollars). Along the northern California coast, two Air Force sergeants were drowned just after midnight while fishing at the mouth of the Klamath River when the tsunami hit. Further south, damages ranged from several million dollars at Noyo Harbor in Mendocino County where 100 boats were damaged and 10-20 were sunk, to about $56 million in San Francisco Bay where docks and boats suffered considerable damage, especially in San Rafael. In Half Moon Bay, Avila and Morro Bay, boats broke loose, were damaged and sunk. Boats and harbor facilities were also damaged in Santa Monica and Los Angeles harbors where total damages reached over $16 million (all in 2017 dollars). There is a pattern here, where nearly all of the significant damage was in or adjacent to ports or harbors, where wave energy can be focused.

The tsunami surge raised water levels 10 feet at the Santa Cruz Small Craft Harbor, which was just being completed. As the water receded, the harbor was drained and boats were left resting on the bottom. The harbor's 35-foot dredge that had been brought in to remove sand from the new harbor was carried out of the harbor by the first surge where it disappeared. Several days later, skin divers, Jack O'Neill and Robert Judd, located the sunken dredge 70 feet off the end of the east jetty in about eight feet of water. The sunken vessel was hauled ashore by a Granite Construction Company caterpillar tractor. A 38-foot fishing boat, the *Big Boy* was damaged as it exited the harbor, perhaps hitting the submerged dredge, and sank quickly with two men jumping overboard, who were then rescued by another boat. Total damage to boats and infrastructure at the harbor reached nearly $6,400,000 (in 2017 dollars). Water came up to the Boardwalk steps with waves described as being eight feet high. A 14-foot wave was reported at Capitola as water overtopped the seawall along the Esplanade in Capitola, which was described at the time as "*a not uncommon happening at high tides*".

At Moss Landing, maximum wave heights of five feet were reported with strong currents in the harbor entrance. Waves 8.5 feet high surged in the bay at Monterey with waves coming in at 20-minute intervals and water elevations reaching a maximum of 7.5 MLLW. At Pacific Grove, maximum water elevations were measured at seven feet above MLLW with maximum wave heights of six feet.

MARCH 11, 2011 JAPAN EARTHQUAKE AND TSUNAMI

The March 11, 2011 Tohoku, Japan 9.0 earthquake, one of the largest in the last century, was accompanied by 23 to 33 feet of seafloor uplift off Japan, which generated a tsunami that propagated both directly onshore and also offshore into the Pacific Ocean. A significant portion of the east coast of Japan was devastated as the tsunami reached maximum elevations of 133 feet above sea level, flowed as far as six miles inland, and flooded about 217 square miles of coastal land. Over 90% of the estimated 19,575 earthquake-related deaths were due directly to the tsunami.

FIGURE 3.15. The intertidal area offshore of Capitola was exposed between the arrival of the individual waves from the 2011 Japan tsunami. View looking west towards the groin and Opal Cliffs.

The waves spread out across the Pacific and produced elevated water levels and moderate damage from Alaska to Chile *(See Figure 3.4)*. Between the Pacific Tsunami Warning Center in Hawaii and the U.S. National Tsunami Warning Center, watches and warnings went out across the Pacific Basin and along the west coast from Alaska to California. With accurate advance warnings, people were notified, evacuations of low-lying areas were carried out, and there was only a single fatality along the entire 1100 mile coast of California – a photographer standing along the northern California shoreline to photograph the incoming waves who didn't take the warning seriously.

FIGURE 3.16. The shoreline west of Cowell's beach was exposed during the 2011 Japan tsunami. (Photo courtesy of Ann Gibbs, United States Geological Survey).

Local observers noted the water receding and returning along the beaches of northern Monterey Bay (*Figures 3.15 and 3.16*). Because of the advance warnings, the Boardwalk, Municipal Wharf, and low-lying streets near the beach front in Santa Cruz were closed. City officials advised about 6,600 people in the tsunami inundation zone to evacuate. Although the warning was advisory and not mandatory, many residents in low-lying areas drove up onto the University campus or even to the crest of Highway 17. At 1,800 feet above sea level, this is playing it "*very*" safe.

The Santa Cruz Small Craft Harbor, as in past tsunamis, again received the brunt of the impact. Shortly before 8 am the first of eight to ten surges swept up the harbor, each about 10 minutes apart. The surges were described as being more like a river than a big wave. Peak height of the waves

FIGURES 3.17a and 3.17b. The March 2011 tsunami from Japan left $28 million in damage to boats and facilities at the Santa Cruz Harbor. (Photos courtesy of the Santa Cruz Port District).

reached five to six feet and waves traveled up the harbor at velocities of up to 15 knots (16.5 miles/hour). Boats were slammed into each other and into some of the docks (*Figures 3.17a and 3.17b*); one 30-foot boat was partially sunk and drifted towards the harbor mouth but was stuck for a while under the Murray Street Bridge. One dock "*just blew up. It buckled and splintered*" according to one observer. When it was all over, about 30 boats had broken free from their dock moorings, 14 sunk and dozens of others were damaged. Of the harbor's 29 docks, 23 sustained significant damage ranging from severe cracking of floats to complete dock destruction. Total losses to harbor facilities, including docks, pilings and other infrastructure, as well as boats, ultimately reached about $28 million. Damage and coastal flooding could have been much worse, however, as the tsunami from 5,000 miles away arrived at low tide. In Monterey, the tide gauge showed four distinct pulses of water, reaching about 2.2 feet above normal sea level (*Figure 3.18*) but no significant damage.

At Moss Landing the tsunami surge did cause five to eight foot waves in the harbor, but with Elkhorn Slough directly behind the entrance

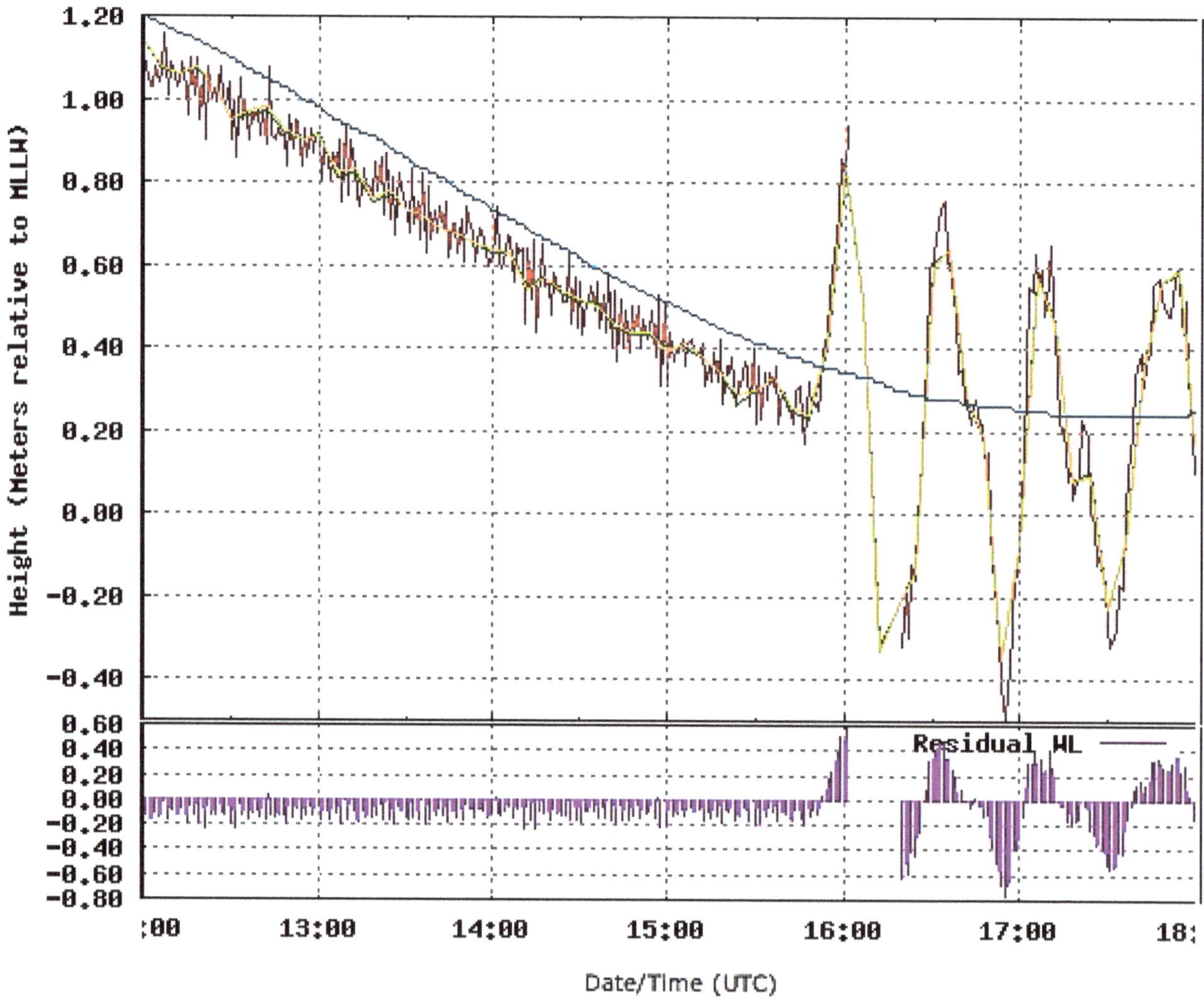

FIGURE 3.18. The record of the 2011 tsunami on the NOAA tide gauge in Monterey. (Graph courtesy of the National Oceanic and Atmospheric Administration).

channel, there was a large area in which to dissipate wave energy. There was little damage to vessels, docks or other parts of the harbor. Subsequent engineering surveys, however, found that 220 timber pilings used to anchor docks and berthing facilities had been damaged from wave scouring during the tsunami. About $1.5 million in federal emergency funds were approved to remove the damaged pilings and replace them with pre-cast concrete pilings.

A student at Moss Landing Marine Laboratories did make some observations during the 2011 tsunami:

"And what did happen in Moss Landing? Those that were here reported that the slough in front of the labs drained, then filled up about 3 feet higher than was normal for the tide, then rapidly drained again as the surge rushed back out. This happened repeatedly, and could also be seen in the harbor. Even though some of our facilities on the sand spit are on much lower ground than the main labs, no tsunami surges breached the land and no flooding occurred. The main lab, up on our hill, was actually used by some of the locals for high ground and refuge from possible tsunami waves."

SOME FINAL THOUGHTS ON TSUNAMIS

The size of a tsunami and its impact can vary widely depending on the magnitude of the earthquake and seafloor displacement, the nature of the offshore bathymetry or bottom topography, the geometry of the shoreline and the coastal topography. Because most tsunamis approaching the coast of California have come from Alaska, Japan, South or Central America source areas, they must pass over many miles of shallow continental shelf before they reach the coastline. As a result, wave energy is significantly reduced, and damage has historically been far less when compared with many other areas around the Pacific basin, such as Japan in 2011.

Although destructive tsunamis are not everyday events in California, they do occur and will certainly occur in the future. Thirteen or fourteen destructive tsunamis have reached the California coast over the past 200 years and about 17 lives have been lost. The tsunamis of 1946, 1964 and 2011 caused the most damage around Monterey Bay and are good indicators of what could happen in the future from a large subduction zone earthquake. The very small number of tsunamis that have had any effect on the Monterey Bay shoreline during historic time, compared to the far greater number of earthquakes and floods, is a good indicator of the overall risk posed by this hazard.

A large landslide or slump in the head of Monterey Submarine Canyon, offshore from Moss Landing, could also generate a modest tsunami, although how frequently such events occur is not known and models indicate that waves would not likely exceed five feet in height at the shoreline. Such an event could be generated by slope instability but more likely from a very large earthquake on the San Andreas Fault or the offshore San Gregorio Fault. Relative to the other natural disasters we face and have experienced in the region, the historical record indicates that the impacts of tsunamis from several very large earthquakes around the Pacific Basin (in 1946, 1964 and 2011) have been relatively minor. While there has been damage to the Santa Cruz Harbor from these events, there hasn't been any significant damage elsewhere around the bay, which should give us some sense of security about future tsunamis.

CHAPTER 4

Coastal Storms, El Niños, Shoreline Erosion and Flooding

Let the waves crash down over me – Logan Kendall

INTRODUCTION

While there are a lot of reasons to live somewhere around Monterey Bay, most of us are probably here, rather than in Merced or Stockton, because of the bay: we want to be near the ocean. We need it in our lives. We like to walk, jog, bike, sit, swim, surf, fish, dive, sail or some other ocean related recreational activity, or maybe all of them if we had the time. And the coastline and waters of the bay allow us all of these opportunities and more. Sure, there is a long wait to get a berth for a boat in the harbor at Santa Cruz, and there are always ten other people taking off on the same wave at Steamer Lane or Pleasure Point, and sometimes you have to look hard for a place to put your blanket on Main Beach, but we are willing to put up with all of these inconveniences. We're happy to pay that price for living in paradise; well, that and being able to afford a house to buy or rent.

For many, however, walking or swimming isn't enough – they want to live right on the edge. They want an unobstructed ocean view or to be able to step off their deck right onto the sand or both. And you can do this around the shoreline of the bay. From De Anza Mobile Home Park next to Natural Bridges State Beach to Pajaro Dunes in Santa Cruz County, and from Del Monte Beach to Carmel, the oceanfront is almost continuously blanketed with homes and there aren't a lot of vacant parcels. Along West Cliff Drive, all the homes but one are on the somewhat safer inland side of the street. Along East Cliff, Opal Cliffs and on Depot Hill, however, the houses are right at the cliff edge, literally... some less than 10 feet (*Figure 4.1*). A few have even dangled over the edge from time to time and been relocated or removed. From Pot Belly Beach to Aptos Seascape, homes have been built not only on the cliff top but also at the foot of the cliff on the beach itself, literally within a good high tide or a stone's throw of the water. If you can fish from your living room window, however, you are too close to the ocean.

The risks to oceanfront development aren't at all apparent in the warm summer months. The beaches are high and wide, the waves are relatively small and the water may even be almost swimmable, and it isn't raining. This is when all of those oceanfront homes are bought and sold. *"Capitola – Iconic home on the beach. Stunning. Price Reduced $4,800,000,"* reads a real estate ad in the Sunday paper in mid-summer, *"Steps to the sand"* (*Figure 4.2*). The next winter's headlines often read: "*Capitola Under a Foot of Water*" or "*Giant Waves Flood Capitola Village*" (*Figure 4.3*). There are some wonderful benefits to living directly on the shoreline, but sooner or later there will also be some very large risks and high costs.

Like earthquakes, our memory of coastal storms and shoreline erosion is fleeting. Unless we suffered damage first hand in the El Niño's of 1978, 1982-83, or 1997-98, the storm damage along the California coast from those storms are ancient history, and lost to the old newspa-

FIGURE 4.1. This apartment complex and these two homes are perched on the edge of an eroding cliff in Capitola. The lower portion of the cliff has been stabilized by the emplacement of a tied-back seawall. (Photo Kenneth and Gabrielle Adelman © 2011).

FIGURE 4.2. The floor level of this pink condominium unit for sale is just a few feet above the beach and all of these units are periodically flooded by high tides and large waves. (Photo courtesy of the *Santa Cruz Sentinel*).

pers. The subsequent summer days or drought years usually erase or smooth over the storm images most people retained. This mental process is sometimes known as collective amnesia.

- *"High tides and waves destroyed bathhouse in Santa Cruz. Downtown Capitola flooded. Venetian Court apartments undercut. High waves wash over 2000 feet of new seawall at Seacliff Beach. Portions of seawall undercut and Beach Drive almost entirely washed away."*

- *"Southerly gale winds and waves swept across Aptos Beach Drive at the Rio del Mar Beach, 15-foot combers smashed against residences. Beach Club severely damaged by waves at Rio del Mar Beach and seawater and sand flooded many of the homes along the beach."*

- *"Southerly winds up to 45 mph with gigantic waves. Rio del Mar, Capitola and Seacliff took the brunt of the waves. At Capitola, waves smashed a beach restaurant and undercut concessions. In Rio del Mar, 25 luxury homes along Beach Drive were damaged by gigantic waves. The Seacliff State Beach camping site was destroyed and restrooms heavily damaged."*

FIGURE 4.3. The Venetian Courts are purported to be the oldest condominiums on the California coast. Built around 1926, they have survived numerous stormy winters and episodes of flooding. (Photo courtesy of Bruce Richmond, 1998, United States Geological Survey).

These news articles should sound familiar to anyone who was living in Santa Cruz County during the winter of 1983. But the dates of the storms described in these stories were, perhaps surprisingly, from February 11-15, 1926; October 27-29, 1950; and February 9-10, 1960. While newer residents often see storms like those of early 1983 as "*acts of god*" or some other misplaced wrath or anger, a look at the storm history recorded in the old newspapers provides a vivid record of these recurring events. And if you spend some time reading through these headlines and stories, it becomes evident that the same geographic areas keep popping up again and again. There are some reasons for this.

While there are certainly benefits to living directly on the shoreline, sooner or later, in all likelihood there will be some very large risks and costs. For many, living on the ocean is a dream come true. For others, it can quickly become a nightmare and an expensive habitat to maintain. Take Joe, for example. In January 1983 he and his family moved into a $350,000 beachfront house literally on the sand near Moran Lake along East Cliff Drive (remember this was 1983 when you could buy a beachfront home for $350,000). Ten days later, during one of the early 1983 storms and simultaneous high tides, the concrete slab the house was built on was undermined and tilted steeply towards the bay. His house slid off the foundation into the oncoming waves. Within minutes nothing remained but a pile of wreckage. Events like this are tragic but, unfortunately, not uncommon.

A dilemma of increasing magnitude has developed in recent years between ocean front development and the inherent geological instability of the shoreline. California now has nearly 40 million people with 68% living in coastal counties and 80% living within an hour of the coast. Yet nearly 90% of the state's 1,100 miles of coast is eroding. It's challenging and expensive to stop the Pacific Ocean from its natural tendencies, although we have spent considerable public and private money attempting to do this, but often on only a very temporary basis.

The natural ongoing processes of beach erosion and cliff retreat were either not recognized, appreciated, or were largely ignored by most developers, builders or homebuyers in the recent past. Within the past four decades the problem has come into clearer focus along virtually the entire state's coastline. The high tides and storm waves during the winter of 1983 inflicted over $250 million (2017 dollars) in damage to oceanfront property. Damage in these storms wasn't restricted to undermined decks and broken windows, 33 ocean front homes were completely destroyed (*Figures 4.4 and 4.5*). Recreational facilities like piers, parking lots, access stairs and bathrooms from San Diego to San Francisco suffered over $66 million in losses (2017 dollars; *Figure 4.6*). Damage to large engineering structures like breakwaters and jetties designed to protect harbors from storm waves reached $50 million.

COASTAL PROCESSES AND HAZARDS

Other than the crater or flank of an active volcano, the shoreline is probably one of the most active geological environments on the planet. Waves, tides, wind, currents, rain, storms, and human activity are constantly changing and reshaping the coastline. The beach comes and goes, cliffs crumble and collapse, seawalls are built, houses appear and sometimes disappear. Sea level is rising globally (about 0.13 inches a year at present) but at any particular location, the coast can be stable, rising or sinking depending upon the regional geologic setting.

FIGURE 4.4. A home along Beach Drive in Rio Del Mar collapsed onto the sand when the high tides and large waves of the 1983 El Niño scoured the sand deep enough to undermine the timber pile foundation.

FIGURE 4.5. Homes along Beach Drive in Rio Del Mar suffered major damage, and failure in some cases, when 1983 El Niño storms battered the protective bulkhead and seawall, scoured the sand off the beach, and removed the support for the timber pilings.

FIGURE 4.6. This timber bulkhead at Seacliff State Beach that supported a recreational vehicle camping area has been damaged or destroyed ten times in 58 years. The 1983 El Niño storms destroyed 700 feet of the 2,600-foot long timber bulkhead that had just been completed the year before.

The existence of a steep sea cliff fronting the ocean such as we see along much of the northern Monterey Bay coastline between Natural Bridges and Rio del Mar indicates that the coastline is being regularly attacked by waves and is eroding. The only question is how rapidly and what risk it poses to the house or hotel, sidewalk or road at the bluff top, or even someone on the beach below.

The presence of beach sand and driftwood along any shoreline indicate that waves have been there in the recent past and will revisit again, perhaps as soon as the next high tide, maybe not until the next winter, but the ocean will return. The simple truth is that the sand and driftwood were delivered by waves and they have a habit of returning regularly, usually at every high tide. It shouldn't come as a surprise to someone who builds or buys a house on the sand to find the waves pounding on their picture window or sliding glass door some stormy winter morning (*Figure 4.7*). The realtor should have disclosed this information, but more often, being "*on the sand*" or "*steps to the beach*" is seen as a strong enticement to a newcomer to the area to buy the house (*Figure 4.8*).

FIGURE 4.7. Waves overtopped this timber bulkhead during the 1983 El Niño storms and shattered the sliding glass doors of this house on the beach at Rio Del Mar.

On the Sand at Seacliff Beach

FIGURE 4.8. A common real estate advertisement for a home "On the Sand". It's hard to miss all of those huge rocks protecting the front of the house. One place you don't want to be investing your life savings is "on the sand." (Image courtesy of the *Santa Cruz Sentinel*).

The two coastal hazards of greatest concern to any oceanfront development around the shoreline of Monterey Bay, whether public or private, are coastal cliff or bluff erosion and flooding by storm waves and extreme high tides.

COASTAL CLIFF AND BLUFF EROSION

Anyone who has spent time watching large storm waves battering the coastline quickly comes to appreciate the tremendous power exerted by an angry ocean (*Figure 4.9*). Even during the calm summer conditions, the 5,000 to 10,000 waves that typically break on the bluff or beach every day have their impact by constantly washing back and forth across the shoreline, grinding away at the rocks. This constant wetting and drying of the cliffs gradually weakens them, just as the water, salt and sun impact anything we leave along the coast or near the ocean for very long, including our cars, our boats, our homes and our skin.

Although this day-to day-activity takes its toll over time, it is generally the large winter storm waves arriving at times of high tides or elevated sea levels that occur during El Niño events that are responsible for most cliff and bluff retreat. The winter waves are usually larger and potentially more destructive, and the winter rains and run-off, as well as groundwater seepage through the cliff rocks, weaken the cliffs making them more susceptible to failure. Another equally important factor, during the winter the beaches are narrow or eroded altogether such that the waves attack the cliffs directly, or at least more often.

FIGURE 4.9. Most coastal erosion and coastal storm damage occur during periods when large waves arrive at times of very high tides. With changing climate we will experience larger waves as well as an increase in the rate of sea-level rise. (Photo: Shmuel Thaler/*Santa Cruz Sentinel* © 2006).

FIGURE 4.10. A small coastal freighter, the *La Feliz*, went on the coastal rocks in 1924 just north of where Natural Bridges State Beach is today. The mast was taken off and used with a block and tackle to salvage the cargo. (Image courtesy of the Santa Cruz Museum of Natural History).

FIGURE 4.11. A photograph taken in 2006 from the same location as Figure 4.10 showing very little change in the configuration of the coastal rocks 82 years later. Note the mast of the *La Feliz* still leaning against the cliff. (Photo: Deepika Shrestha Ross © 2006).

CLIFF EROSION ALONG THE SANTA CRUZ COUNTY COASTLINE

Where the rocks making up the sea cliffs are soft or weak, wave energy is high and the waves regularly attack the cliffs, we can expect high rates of coastal retreat. Along the coastline of northern Monterey Bay both the type of rocks exposed in the cliffs and the wave energy vary from place to place. In general the mudstone, which makes up the north coast of Santa Cruz County, and most of West Cliff (at least south to approximately Almar Avenue) is fairly hard and doesn't yield readily to wave action. Where average long-term cliff retreat rates have been measured, they are generally less than six inches per year.

Just west of Natural Bridges State Beach, and directly in front of the Seymour Marine Discovery Center at the University's Coastal Science Campus, is a wooden ship's mast leaning up against the cliff looking like an old telephone pole. On the night of October 1, 1924, the combination of high seas and a course too close to the shoreline put the *La Feliz,* a small coastal freighter, on the rocks directly in front of where the Seymour Center stands today (*Figure 4.10*). The 100-ton vessel was carrying canned sardines from Monterey to San Francisco when she was wrecked. Local residents drove out onto the terrace, lit up the vessel with their headlights and rescued the crew at night. The mast was removed from the ship the next day, leaned up against the cliff, and with a block and tackle was used to salvage the cargo of canned fish. Ninety-four years later, the mast is amazingly still there, indicating that the cliff at this location has undergone almost no change or erosion in this time (*Figure 4.11*). Just 2,000 feet east, however, over the same approximate time period, two of the three arches that gave Natural Bridges State Beach its name have collapsed (*Figure 4.12 a, b, c*). Natural arches and bridges along West Cliff and along the north coast of Santa Cruz County continue to form and collapse, but in general these mudstone cliffs erode at average long-term rates of only a few inches per year. The erosional process

FIGURES 4.12 a, b, c. Natural Bridges State Beach is home to what has probably been the best known and most photographed natural arches that are common along West Cliff Drive in Santa Cruz. This sequence of photographs taken in ~1890, 1972 and 2006 show the progressive collapse of two of the three natural bridges. (1890 photo courtesy of the University of California Santa Cruz Special Collections, and 1972 photo courtesy of Rogers E. Johnson).

is episodic, however, with an arch or cave collapsing and moving the cliff edge back a dozen feet or more, and then not much may happen for another 10 or 20 years.

Along Opal Cliffs, on Capitola's Depot Hill, and through Seacliff and Rio del Mar, the cliffs consist of much weaker sandstone, siltstone and mudstone of the Purisima Formation, which is far less stable and erodes more readily. Historically, earlier homebuilders avoided these cliff top areas. But over the years, ocean views and ocean front properties became more sought after and houses encroached closer and closer to the edge (*Figure 4.13*).

South of Rio del Mar, at Aptos Seascape, La Selva Beach, Place de Mer, Sand Dollar Beach, and Sunset Beach, homes have been built on top of and on the edge of relatively loose and often weak ancient sand dunes. These weak bluffs are very susceptible to failure during intense rainfall, earthquakes, or when under wave attack. Fortunately, wide sandy beaches front these developments most of the time so wave erosion is not a regular threat. The 1989 earthquake, however, and the major rainstorms of early 1982 did take their toll on the sandy bluffs (*Figure 4.14*).

The sand dunes at the southern end of Santa Cruz County do present some concerns. The largest development in the area, Pajaro Dunes, consists of 396 condominiums, 24 town houses, and 145 single-family dwellings immediately north of the mouth of the Pajaro River. All of the units were built on active sand dunes, with many of the homes, townhouses and condominiums built right on the frontal dune, directly above the beach. Centuries of experience living with dunes in places like northern Europe and the East Coast of the United States has made it clear that

FIGURE 4.13 Continuing cliff erosion has brought homes built along Opal Cliffs Drive in Santa Cruz closer and closer to the edge. While some homeowners have constructed seawalls and placed rip-rap, others have no protection. (Photo: Kenneth and Gabrielle Adelman © 2015)

the first line of dunes, those closest to the beach, is a moving buffer zone and not the place to build anything permanent; and the inner Monterey Bay shoreline is no exception. The history of the Pajaro Dunes area over the past 50-75 years, which is evident in the old aerial photographs, is one of dune erosion or retreat during particularly stormy or strong El Niño winters, followed by gradual rebuilding or accretion of the dunes as beach sand blows onshore. So while there doesn't appear to have been significant net retreat of the shoreline, the advance and retreat of the dunes may move the front edge of the dunes 40 or 50 feet during a single winter. Unfortunately the homes and condominiums don't move with the dunes.

Since development of the dunes started in 1969, four major El Niño winters (1978, 1980, 1982-83 and 1997-98) have brought large waves from the west and southwest, combined with elevated sea levels, and seriously eroded the frontal dunes. The January 1983 storms and high tides cut back the dunes up to 40 feet and left a near vertical cut measuring 15-18 feet high that came right up to the foundations of some of the homes (*Figure 4.15*). Only the emergency emplacement of thousands of tons of rocks along the base of the eroded dunes saved these expensive homes from disaster. Following the winter storm season, a permanent revetment was built along the seaward frontage of the entire development at a cost of several million dollars. Although the rock has provided some protection, by the time the 1997-98 El Niño arrived, much of the well-planned revetment was scattered across the beach and had to be rebuilt. Any resemblance to the original natural dune environment has long-since disappeared. A rising sea level will be a challenge for these homeowners.

FIGURE 4.14. During the intense and prolonged rainfall of January 1982, bluff failure along Beach Drive in Rio del Mar led to damage and destruction of several houses at the base of the bluff.

FIGURE 4.15. The simultaneous occurrence of high tides and large storms waves during the El Niño winter of 1983 led to severe erosion of the frontal dunes at the Pajaro Dunes development.

the Loma Prieta earthquake in 1989 was smaller, its epicenter was far closer. The severe shaking again produced extensive liquefaction in the Moss Landing area, which unfortunately led to the destruction of the laboratories *(See Figure 2.29)*. It took nearly a decade to obtain the funds and permits to rebuild the facilities a short distance inland on a stable hilltop site.

Just north of the Salinas River mouth, the Monterey Dunes Colony is built well back from the frontal dune and hasn't experienced any significant threats from shoreline erosion. Either the architects for this project, or the planners for Monterey County, or both, made a wise decision to avoid the frontal dune and the homes are safer as a result.

High dunes, some still actively migrating and some now vegetated and stable, back the six miles of shoreline south of the Salinas River. This area is essentially undeveloped except for a sand mining operation that has been pumping the unique coarse-grained and amber-colored sand out of a back beach pond for a century or more for commercial purposes. After decades of efforts to terminate this sand mining because of the impacts on shoreline erosion, agreement was finally reached in 2017 between the California Coastal Commission, the State Lands Commission, the city of Marina, and CEMEX, the owner and operator of the sand mine, to terminate operations at the end of 2020. Several major downcoast developments suffered from the erosion induced by the removal

EROSION ALONG THE MONTEREY COUNTY COASTLINE

Crossing the Pajaro River we enter Monterey County, where other than the industrial, research and fishing industry activity centered at Moss Landing, ocean front development along the shoreline is much less intensive. The Moss Landing area has a long and colorful history extending back to the 1850s that includes at one time or another: salt ponds, whaling stations, a marine biological laboratory, a magnesium extraction plant (Kaiser Refractories), a large thermoelectric power plant, Moss Landing Marine Laboratories, the Monterey Bay Aquarium Research Institute, and a fishing harbor, among other things.

The sand spit, also known as "*the island*", where the original Moss Landing Marine Laboratories were built, is a site that experienced well-documented liquefaction and major ground failure during the great 1906 San Francisco earthquake *(See Figures 2.22 and 2.23)*. This history was disregarded in the 1960s when the State University Marine Labs were first constructed. Although

of about 250,000 cubic yards of sand annually (about 25,000 dump truck loads) for decades, however.

FIGURE 4.16. Continuing retreat of the sandy bluffs fronting Stilwell Hall at the former Ft. Ord led to repeated dumping of rock and broken concrete in an effort to slow the erosion (2003). Note that while a beach exists to either side of the Hall, there is no beach in front of the armor. (Photo: Kenneth and Gabrielle Adelman © 2003).

FIGURE 4.17. Site of the former Stilwell Hall in 2005 after building and rock were removed. With continuing retreat the beach has now returned. (Photo: Kenneth and Gabrielle Adelman © 2005).

The former Soldiers Club at Fort Ord, also known as Stilwell Hall, was an entertainment center where many World War II soldiers spent their last evenings before being shipped overseas. Bluff erosion rates in front of Stilwell Hall averaged six feet or more per year for decades. Sand dunes erode much more rapidly than rock cliffs and the dunes beneath Stilwell Hall were no match for winter waves and high tides. The structure was closed some years ago because of its precarious bluff-top location. Over the years, in an effort to protect the structure, the Army regularly dumped rocks and old concrete slabs over the bluff to try and halt or slow the ongoing retreat (*Figure 4.16*). Winter waves, however, broke over the rocks and undermined and eroded the loose sand. By early 1984 the edge of the eroding bluff was within 15 to 20 feet of the structure. Although the Army considered different approaches for the historic building, they ultimately decided that it wasn't cost-effective to try and save it. When Fort Ord was closed in 1994 and the bluff beneath Stilwell Hall was becoming a peninsula, the decision was made to demolish the historic structure and remove the rock and concrete that had been dumped on the beach. The demolition and beach cleanup was completed in 2004 and walking on the beach today, there is no indication that there was ever any armor on the shoreline (*Figure 4.17*).

A half-mile south of the hotel, a large 172-unit apartment complex was built on the dunes between 1972 and 1974, where average long-term retreat of the dunes was about 1-2 feet annually. As with the Monterey Beach Hotel to the north, the apartment complex was approved by the City of Monterey prior to the establishment of the California Coastal Commission. The oceanfront units were first seriously threatened by the coincidence of high tides and large storm waves combined with the elevated sea levels during the 1982-83 El Niño. Wave erosion broke a water line, which had to be rerouted, and threatened the sewer and electrical lines. Continuing dune erosion by January 1984 had approached to within 14 feet of the shallow timber pilings supporting the apartments (*Figure 4.19*). Five thousand tons of rock were brought in to provide emergency protection for the frontal units, but because the apartments were built right to the property line, the rock had to be placed on beach belonging to the city of Monterey.

FIGURE 4.18. Failure of the south end of the seawall protecting the Monterey Beach Hotel led to emergency placement of rock followed by reconstruction of the wall.

Four miles south of Fort Ord, a hotel was constructed in 1968 on the back beach. This hotel was approved prior to the establishment of the California Coastal Commission, which now takes a much harder look at any oceanfront construction. To protect the building, a concrete panel seawall was built completely across the front and along both sides of the new hotel. Wave impact and overtopping during the severe El Niño storms of 1983 surged through the joints between the concrete panels, leading to some loss of fill behind the wall. Large waves combined with high tides in 2002 destroyed much of the south side of the seawall and began to erode a parking lot, which required emergency placement of rock (*Figure 4.18*). The dune edge has now migrated significantly landward of the front of the seawall, so that the hotel is becoming a peninsula during most high tides, leaving no beach access in front of the structure.

After an extensive Environmental Impact Report (EIR) process, and because of the loss of public access along the shoreline at high tides, the city required that the emergency rock be removed. While one option analyzed in the EIR was to remove the threatened front units, this was turned down in favor of re-supporting these structures with a series of 50 to 55-foot-deep concrete piers connected by concrete grade beams (*Figure 4.20*). One factor that complicated any decision-making was that the original apartment units had been converted to condominiums with over 150 different owners.

Although the sandy beach fronting the development narrows and widens seasonally as do almost all beaches, the long-term pattern has been one of continuing retreat of the outer edge of the dunes at a rate of one to two feet per year. The 1997-98 El Niño storms and elevated sea levels caused additional retreat. By 2002, more emergency rock had to be emplaced to prevent undermining of additional landward units. Another EIR was required and again the City of Monterey required that the rocks be removed from the public beach and that a more permanent solution be found. While a number of different options were evaluated in the EIR, along with costs and environmental impacts (again, including demolition of the frontal units), the solution ultimately selected and approved by the California Coastal Commission because of the threat to the survival of the front row of condominiums was a massive concrete seawall. The consultants described this structure at the time it was proposed as having the appearance of an eroded

FIGURE 4.19. Severe erosion of the dunes threatened the Ocean Harbor House development, which was originally only supported on shallow timber piles (1984).

FIGURE 4.20 Following placement of temporary rip-rap while an Environmental Impact Report was completed to evaluate different long-term solutions for Ocean Harbor House, the rip-rap was required to be removed. Fifty-five foot deep concrete caissons connected with concrete grade beams (the cross-members connecting the caissons) were then emplaced to support the front units (1998).

sand dune. You can decide for yourself if they achieved this objective or claim *(See Figure 1.3).* With continued bluff retreat at one to two feet per year, it was recognized that there would be no beach access along the shoreline at high tide and that the condominiums will ultimately become an armored peninsula.

HOW FAST ARE THE BLUFFS ERODING?

By looking at historical maps and old photographs, particularly vertical aerial photographs, and carefully measuring the distances from recognizable features such as streets or houses to the cliff edge, we can get a relatively accurate picture of how fast a sea cliff or coastal bluff is eroding. This is useful information to have before buying a clifftop home, but isn't usually readily available unless you know where to look or who to ask.

FIGURE 4.21. Grand Avenue along the top of Depot Hill in Capitola (c. 1890) was originally a wide public street with two rows of Monterey Pine trees and a sidewalk separating it from the cliff. (Photo courtesy of the Capitola History Museum).

Natural Bridges State Beach was originally named for the three arches or bridges, which have now been reduced to one through erosion and collapse over the past century *(See Figures 4.12 a, b, c).* The collapse of bridges and arches is one of the more spectacular and visible examples of sea cliff erosion, a process that is not usually obvious but has been continuing along the cliffs of northern Monterey Bay for thousands of years.

Some of the most dramatic erosion in Santa Cruz County occurs along the 75 to 95-foot high cliffs of Depot Hill between Capitola and New Brighton State Beach. The rocks are particularly weak and there is seldom a protective beach of any width such that the waves attack the base of the cliffs virtually daily. Long-term erosion rates here over the past 50 or 75 years average about a foot annually. Grand Avenue along the cliff top used to be continuous with a two rows of pine trees on the seaward edge separated by a pathway with benches known as Lover's Lane (*Figure 4.21*). The trees are long gone and much of the road has now been reduced to a narrow walkway or nothing at all. Two houses have been relocated over the years and six clifftop apartments were demolished after they were further undermined by cliff failure during the 1989 earthquake (*Figure 4.22).* While numerous schemes have been proposed to protect these high vertical cliffs from further erosion over the past several decades, they are all very expensive and because of the height of the cliffs and the weak rocks themselves, do not guarantee that cliff collapse will come to a complete halt. Several individual cliff top property owners have been successful, at least over the short term, in stabilizing the underlying rocks essentially using rock bolts to retain the cliff materials (*Figure 4.23*), although this is very expensive.

FIGURE 4.22. The Crest Apartments on Depot Hill were built almost at the cliff edge in the 1960s. Continuing erosion and then cliff failure during the Loma Prieta earthquake led to loss of support for these concrete caissons followed by demolition of six units.

While most bluff top homeowners between New Brighton Beach and Aptos Seascape felt somewhat secure knowing that they were safe from wave attack because of recently constructed and expensive seawalls, the effects of the 1989 earthquake surprised them. With an epicenter only about ten miles away, shaking of the bluffs was severe. The immediate effect was collapse of the loose material on the bluff face and cracking of the bluff top materials underlying the homes. Bluff failure was nearly continuous from Capitola's Depot Hill to Sunset Beach. Six apartments on Depot Hill suffered foundation failure and had to be demolished. Cliff top and beach level homes at Pot Belly Beach, along Beach Drive, at Aptos Seascape, Place de Mer and Sunset Beach were all affected *(See Figure 2.37)*. The bluff debris, which slumped downslope, damaged homes at the base of the bluff and blocked access in several places for days. As discussed in the earthquake chapter, the few coastal accounts of earlier large earthquakes indicated that bluff failure was common. The last major earthquake was in 1906 so this wasn't a process that had been recently observed or recognized as a coastal hazard.

Construction on oceanfront sand dunes doesn't present the same hazard as cliff top construction, but the history of Pajaro Dunes indicates that this is not always the safest place to build either. Sand dunes are constantly changing along with many other coastal landforms. When there is an ample supply of sand, strong onshore winds, and a low relief back beach area where the sand can accumulate, as was the case around the edges of inner Monterey Bay in pre-historic times, dunes may form and move inland until stabilized by vegetation, the wind dies, or there isn't an adequate supply of sand. The active sand dunes around inner Monterey Bay extend from Sunset Beach to Monterey and include areas like Ft. Ord, Sand City, Marina and Seaside.

Dunes are constructed of sand blowing onshore from the beach and will grow in size when ample beach sand is available, but will be eroded during severe winter storms when large waves scour the beach and reach the dunes. The history of sand dunes anywhere is one of advance and retreat or accumulation and erosion. Pajaro Dunes is no exception, although the planners, architects and engineers who designed the Pajaro Dunes project didn't understand this natural process of change and evolution. The historic aerial photographs of the Pajaro Dunes area, which extend back over 90 years, indicate that the south end of the development, an area now covered with the Pelican Point condominiums and a number of individual homes, had been completely washed over by waves in the late 1920s (*Figure 4.24*). Elsewhere along this dune front, photographs indicate advance and retreat of up to 30 feet has taken place over the years.

FIGURE 4.23. This section of Depot Hill, about a half a mile east of the Crest Apartments, has been stabilized, at least for the short term, by using a tied-back shotcrete wall. The ends of some of the tie back rods are now exposed.

A resort was built on the back dune area just south of Beach Road (which connects Watsonville to the Pajaro Dunes development) in about 1881, and consisted of a hotel and cottages. A pier in the vicinity had been constructed about 14 years earlier, but was destroyed by a storm in 1904. In 1911 a new 1700-foot long wharf was constructed just north of the present Pajaro Dunes condominium development. The next year the new wharf was damaged by a large storm. "*On the beach, the waves dashed up to the Casino Building* (part of Camp Goodall) *on top of the sand dunes.*" This storm was described at that time (*Watsonville Pajaronian*, October 1912) as "*the heaviest sea in the history of Monterey Bay" and "waves threatened to overflow the sand dunes on the beach*". Two months later in December 1912 another large storm rolled into Monterey Bay and "*huge breakers rolled over Port Watsonville flooding Calpaco*" (a short-lived tent city resort with boardwalks and running water to each tent) located nearby in the dune field, which took its name from the parent company, The California Pacific Company, who were real estate promoters.

In 1969 as the Pajaro Dunes development was in the initial stages of construction, storm waves attacked the dunes. Twelve of the newly created ocean front lots were severely eroded; old automobile bodies were brought in and placed at the base of the eroded dune for protection from further wave impact. Nine years later in the 1978 winter, the first large El Niño in several decades brought elevated sea levels to the central coast. Large waves from the southwest combined with high tides eroded the beach back to the dunes along a frontage of over a mile and erosion threatened a number of the homes. Empty steel barrels, large concrete blocks and sandbags were frantically emplaced by residents in order to halt the erosion.

FIGURE 4.24. The mouth of the Pajaro River in 1928. The unvegetated area of white beach sand upcoast (left) of the river mouth in this photograph has been recently overwashed by storm waves. The left portion of this overwashed area was the site of construction of the Pelican Point townhouses about 40 years later. (Photo courtesy of the University of California Santa Cruz Map Library).

The dunes had nearly recovered when the early 1983 El Niño storm waves and high tides arrived. Portions of the dunes were cut back 40 feet virtually overnight, which soon threatened 17 homes and 25 condominiums. Subsequent storms worsened the situation and over a million dollars in emergency rock was soon brought in to save the development. Ultimately a permanent revetment was constructed along the entire oceanfront of Pajaro Dunes at an additional cost of several million dollars (*Figure 4.25*).

BEACH CONSTRUCTION AND FLOODING

Taking a casual walk along the shoreline of northern Monterey Bay, it may come as a surprise to find dozens of homes actually built on the beach, right out there on the sand. Somewhat difficult to believe, but part of our historical, pre-Coastal Commission development pattern around northern Monterey Bay was construction directly on the shoreline. Out of town visitors have often asked: "*Why would you build a house on the beach?*" Yet for a distance of over 2.5 miles, from Pot Belly Beach on the north to Aptos Seascape on the south, a state beach recreational vehicle campground, parking lots, restrooms, a sewer line and dozens of very expensive private homes have all been built on the back beach. The historic record shows repeated flooding of these areas during high tides and storms waves, particularly during large El Niño events when sea levels may be

FIGURE 4.25. This section of single family homes in the Pajaro Dunes development was built on the outer edge of the frontal dune. Following major dune erosion during the 1983 winter storms a mile-long rock revetment was constructed.

raised a foot or more above normal high tides for several months. The presence of beach sand and large stranded driftwood logs is testimony to the level that waves in the past have reached. The back beach area is analogous to a river's flood plain. The question is not if it will be covered by seawater, but when, how often, and how deep?

The entire shoreline from New Brighton State Beach to Aptos Seascape was affected by the high tides and El Niño storm waves of 1978, 1982-3, 1997-98 as well as a number of very high King tides. Homes along Las Olas Drive were seriously damaged during the 1978 storms as the 75-foot wide beach was eroded allowing the storm waves to scour the sand from beneath the foundations. One house partially collapsed while others suffered broken plate glass windows, damaged patios, decks and stairways (*Figure 4.26*). Las Olas means the "*waves*" in Spanish. Who named this road and were they trying to tell us something? Malibu, interestingly, which is facing some shoreline erosion problems of its own, has a Sea Level Drive.

The most damaging storms and high tides in several decades struck the central coast repeatedly during the first three months of the 1983 El Niño. Along Beach Drive in Rio del Mar, the Seacliff State Beach parking lot was undermined and collapsed as waves and driftwood logs destroyed much of the protective and recently rebuilt timber bulkhead. A group of 26 homes, all built a few feet above beach level on the seaward side of Beach Drive, extend along the beach south of the parking lot. These homes were protected by a variety of structures including large concrete blocks, timber seawalls, and a concrete bulkhead consisting of sections of an old highway bridge. The combination of elevated sea levels from a warmer El Niño ocean and large waves arriving coincident with high tides destroyed or damaged virtually every seawall or protective structure. The weaker walls, typically timber, were destroyed first as the waves battered them with large redwood logs carried down from the rivers and creeks by flood flows. As the weakest links in the seawall were broken, the waves removed the sand from behind the adjacent structures, which led to total loss of seawall support. With the protection lost, waves continued to scour the sand from the beach, ultimately undermining the shallow pilings supporting several homes allowing them to collapse onto the beach *(See Figures 4.4 and 4.5)*. Adjacent houses lost plate glass windows, sliding glass doors *(See Figure 4.7)*, decks, and beach stairways.

FIGURE 4.26. Severe beach erosion during the El Niño winter of 1978 undermined these homes built on Pot Belly Beach between New Brighton and Seacliff State Beaches.

Seacliff State Beach provides one of the best examples of what we could learn from history if we only took the time to study it. The original Seacliff private development was laid out on the bluff top in the 1920s and included an access road to the beach and a seawall. Storms in 1926, 1927 and 1931 destroyed or partially destroyed the first private seawall, a bathing pavilion and also a concession building. These events should have provided ample warning to the state before they bought the property in 1934 and turned it into a beach park for camping and picnicking, and later for RVs after they were invented.

This northern corner of Monterey Bay is partially protected from the impact of the dominant waves typically arriving from the northwest, but has repeatedly been damaged or destroyed by waves approaching directly from the west or southwest during El Niño years. Seven times in 68 years or about every ten years, seawalls or bulkheads protecting the Seacliff area have been damaged or destroyed. After extensive damage in 1939 and 1940, the bulkhead was rebuilt. A year later storms during the winter of 1941 essentially destroyed it again (*Figure 4.27*). Following extensive damage to a timber piling and wooden bulkhead and the recreational vehicle campground in 1978 and again in 1980, a new 2,700-foot long timber bulkhead, identical to the one just destroyed, was reconstructed by the State Department of Parks and Recreation along with the recreational vehicle campground at a cost of $1,700,000 (over $5 million in 2017 dollars, or $1,850/foot). This new structure, completed in the fall of 1982, was intended to last 20 years. In late January 1983, within two months of its completion and dedication, elevated sea levels from a large El Niño, combined with high tides and storm waves from the southwest battered and overtopped the bulkhead. Large logs acted like battering rams and destroyed over 700 feet of the brand new structure *(See Figure 4.6)*. Logs, sand and debris were carried over and through the bulkhead to the base of the sea cliff. The parking lot, utilities, and RV campground were extensively damaged with total losses placed at $740,000 ($1.8 million in 2017 dollars, one-third of the construction cost of the just completed bulkhead). Residents of Las Olas Drive couldn't reach their homes for nearly a week as their access passes through the state park.

FIGURE 4.27. Seacliff State Beach in January 1941. The pilings to the right are the remains of a timber bulkhead built in 1940 and destroyed during the winter of 1940-41. Structure to the left is the remains of a seawall constructed in 1927 and destroyed the following winter. (Photo courtesy of the Army Corps of Engineers).

FLUCTUATING CLIMATES AND STORM IMPACTS

There is a long, colorful and frightening coastal storm history for Monterey Bay that has been recorded in the local newspapers.

It may be surprising for summer visitors or new residents looking for an oceanfront home to read how many times in the past major storms have attacked the shoreline of the bay and how often the structures we have built along its edges have been overtopped, damaged or destroyed (the seawall at Seacliff for example, or the Esplanade at Capitola). While we cannot predict with any certainty when the next El Niño or major storm is coming, there are some things we have learned over the past several decades.

Oceanographers, meteorologists and fishery scientists have now recognized that the climate over the Pacific Ocean changes over cycles lasting several decades, known as Pacific Decadal Oscillations (PDO). Through years of ocean observations, from ships and satellites, from at sea and along the shore, we have discovered that there are major shifts in ocean temperature and climate. These changes affect not only our storm frequency and intensity, but our rainfall, and can also have big impacts on the abundance and distribution of marine life. These shifts or cycles typically last several decades. Changes in surface water temperatures over large areas affect atmospheric pressures and this, in turn, affects wind patterns. The storm climate, water temperatures, intensity of coastal upwelling, and availability of nutrients all fluctuate over cycles that may last 20 or 30 years.

During warmer PDO cycles, surface waters have elevated temperatures, evaporation is greater, and rainfall is higher along the west coast, followed often by floods and landslides or mudflows. In addition, however, these warmer cycles are also periods when El Niño events are more frequent and stronger, which bring elevated sea levels, more severe coastal storms from the west and southwest, all of which are more damaging to the Monterey Bay region.

Looking back at the historic record of those storms that have been the most damaging to coastal development around the shoreline of the northern Bay (East Cliff Drive, Capitola, Rio Del Mar, Seacliff, and even as far south as Pajaro Dunes), the newspaper accounts almost always describe a storm from the southwest. As discussed earlier, the northern part of Monterey Bay is somewhat protected from the most common waves arriving from the North Pacific or Gulf of Alaska, losing considerable energy as they bend around or are refracted into the bay. El Niño driven storms from the west or southwest, however, strike the East Cliff to Pajaro Dunes area more directly with little energy loss and produce greater damage and destruction. The Monterey area, on the other hand, is tucked in behind the Monterey Peninsula and is more or less sheltered or protected from those waves coming out of the west or southwest. Damage to boats anchored in the Monterey area and to the wharf and pier has occurred historically when waves are coming from the north or northwest, which hit this area with little loss of energy.

From about the mid-1920s to the mid-1940s, the California coast was in a warm Pacific Decadal Oscillation cycle, generally characterized by more severe weather, higher rainfalls, and more frequent and damaging coastal storms. Large El Niño events were common. In the mid-1940s, however, this large Pacific climate system transitioned to a three-decade long, cool PDO interval, with La Niña events dominating. Rainfall was generally lower and there were fewer damaging coastal storms with large waves. This approximately 30-year period, following World War II, was also the time when California's population exploded with new arrivals, increasing 250% from 8,750,000 in 1945 to about 22,000,000 in 1976. New residents flocked to coastal counties where employment opportunities were the best and much of the coastline was developed. Bluffs, beaches and dunes were subdivided and houses were built in places that must have looked fine at the time to the planners, developers and mortgage bankers.

A surprise came in the mid to late-1970s when the climate shifted again back to a warm, El Niño-dominated, PDO cycle. The offshore waters warmed, the rains came, the coastal storms were more frequent and severe, and along with elevated sea levels and high tides, all of that new development and the homeowners were in for some unfortunate surprises. The winters of 1977-78, 1982-83 and 1997-98 stand out as particularly severe El Niño years with considerable damage and destruction around the shoreline of the bay.

THE HISTORY OF COASTAL STORMS AND DAMAGE AROUND MONTEREY BAY

The individual years of major coastal storm damage around the shoreline of Monterey Bay over the last 150 years of reasonably accurate reporting are summarized below in a historic chronology with storm events and their impacts drawn primarily from the historic newspaper files (mainly the *Santa Cruz Sentinel*, *Watsonville Pajaronian,* and *Monterey Herald*). In some cases the language and names of the time are used. What is evident is the frequency of coastal storm damage, and also how different areas are impacted at opposite ends of the bay depending upon whether the storm came from the southwest or northwest. It is also interesting how often the newspaper accounts describe a number of different storms as the worst or most violent storm to ever hit the area. There is no way to assess the accuracy of some of these newspaper descriptions, which often appear to have been exaggerated to some degree.

NOVEMBER 25, 1865

Five hundred feet of the Soquel Landing (now Capitola) Wharf were lost and the remaining 600 feet were deemed "*useless*". Pilings were deposited in a potato field beyond the beach and wharf damage was listed as $6,000. Waves described as "*mountain high*".

DECEMBER 14, 1867

Storm damaged wharves in Aptos and at Watsonville, but no account of damage at Soquel Landing (Capitola).

DECEMBER 23, 1871

A strong southeast gale brought floodwaters nearly 12 feet above the average high tide at the Water Street bridge on the San Lorenzo River.

FEBRUARY 8, 1892

High tides washed the yacht *Petrel* ashore at Capitola. Beach concessions damaged and swimmers endangered. It is difficult to imagine swimmers in the water of Monterey Bay in February.

MARCH 13-14, 1905

Headlines stated: "*The Worst Storm for 27 Years Visits Monterey Bay*", and from all newspaper accounts, this seems to have been true. Five-hundred and twenty-four feet of the Watsonville Transportation Company's new wharf were swept away together with a pile driver at Port Rogers (which was later to become Port Watsonville, being located at what is now Sunset State Beach). Waves from the southwest carried the pilings from the wharf all the way north to Leonard's Station (just north of present day La Selva Beach).

Damage was widespread in Capitola. "*Capitola paid for its advantageous location on the sea beach Sunday night with broken porches, flooded cottages and general wreckage of many of its summer homes. The hotel suffered especially from the storm. About four feet of water stood in the hotel on Monday morning. The bandstand and the hotel porch have been swept away, and the windows torn out on one side of the main office and club room, which face the beach. Sidewalks were washed against the cottages and for a time created a fear that the whole summer settlement would be swept away.*"

Waves swept across Twin Lakes Beach and washed out the approaches to the new streetcar trestle (*Figure 4.28*). Waves washed over Seabright Beach and tore off the porch of the Seabright bathhouse. Cowell's Wharf and the railroad wharf lost a number of piles. The end of the Casino Pier was broken off and its steps carried away. The Bertha Leibbrandt bathhouse on Main Beach was washed off its foundation and almost demolished.

FIGURE 4.28. A streetcar line connected Santa Cruz to Capitola between 1905 and 1926 and crossed Twin Lakes Beach on an elevated trestle, which met the bluffs at the correct elevation to continue down 12th Avenue. See Figure 4.30. (Photo courtesy of the Santa Cruz City County Library System).

NOVEMBER 22, 1910

Monterey Bay was very rough and surf was running high. No ships were able to enter or leave Monterey Harbor.

FIGURE 4.29. High tides and storm waves during the winter of 1912 reached the steps of the Santa Cruz Beach Boardwalk. (Photo courtesy of the Santa Cruz Beach Boardwalk Archives).

OCTOBER 4-11, 1912

A heavy swell and strong northwest wind. Wharves at Monterey damaged and a number of boats were beached. A small-boat wharf was washed away and waves washed over the boardwalk near the railroad depot in Monterey. Waves washed up onto the steps of the Santa Cruz Boardwalk and broke away some of the underlying boards *(Figure 4.29*). Waves also eroded the bluff at the end of 12th Avenue and left a portion of the Santa Cruz to Capitola streetcar track hanging in the air (*Figure 4.30*).

NOVEMBER 27, 1913

A severe storm with "*monster*" waves arriving at high tide washed across the beach at Capitola to the Esplanade and spread "*clear to the railroad tracks*". Water reached the Hihn Superintendent's Building (at Capitola and Monterey Avenues). The huge waves destroyed about 200 feet of the middle of Hihn's Wharf stranding a fisherman, Alberto Gibelli, on the outer end (*Figure 4.31*). He was rescued four hours later when another fisherman, Giacomo Stagnaro, who had run his boat over from Santa Cruz, threw a rope and life preserver to him. Alberto put on the life preserver, tied the rope around his arms, took a deep breath and jumped off the pier into the ocean. Rescuers on the shore end of the pier pulled him to safety.

FIGURE 4.30. Cliff erosion during a winter storm in 1912 left a portion of the streetcar line shown in Figure 4.29 at the end of 12th Avenue suspended in air. (Courtesy Santa Cruz Public Libraries).

APRIL 29-30, 1915

Waves coming from the northwest inflicted considerable damage in Monterey but little was recorded for the northern portion of the bay. Damages included: the city wharf (now Fisherman's Wharf) was buckled and pilings were lost; fish sheds were blown off the wharf by strong winds; as many as 100 boats of all sizes, mostly small, were washed ashore with many damaged; the railroad yards were littered by debris thrown up by breakers, and a boat was sunk at Point Lobos.

FIGURE 4.31. A severe storm in November 1913 destroyed the mid-section of the wharf at Capitola, stranding a fisherman on the outer end. (Phot courtesy of the Capitola History Museum).

JANUARY 27, 1916

No damage was reported on the Monterey Peninsula but Moss Landing suffered damage including the destruction of a sturdy steamship pier.

NOVEMBER 26-27, 1919

The Thanksgiving night storm of 1919 may have been the best remembered by the old-timers of the Monterey area as 93 boats were wrecked, including over half of the entire fishing fleet, thus eliminating the livelihood for a great many people.

NOVEMBER 29-DECEMBER 1, 1923

A strong gale from the northwest swept 15 boats ashore at Monterey and beached a freighter, the *Robin Gray*, at Santa Cruz. Heavy seas offshore.

FIGURE 4.32. Large waves combined with high tides in February of 1926 washed into downtown Capitola and flooded the city with a foot of seawater. (Photo courtesy of the Capitola History Museum).

FEBRUARY 13, 1926

Headlines: "*High Waves do Damage to Beaches*". Never in the history of Capitola, according to its oldest residents, has such havoc been wrought to the beaches and buildings of Capitola than during "*The Famous Lincoln's Birthday Storm*" of February 12,13,14. At 10:10 a.m. when the highest tide of the season arrived, a huge 20-foot wave broke. The two bandstands collapsed almost simultaneously, and the Capitola bathhouse and boathouses were washed away by the waves along with beach concessions. High tides reached a block up the main streets and downtown Capitola was under a foot of water (*Figure 4.32*). The Venetian Court apartments, which had just been completed and replaced the old Italian fishing village, were flooded and their seawall promenade was broken. Wave run-up reached the second floor of the Hotel Capitola (*Figure 4.33*).

FIGURE 4.33. The 1926 storm waves overtopped the seawall protecting the old Hotel Capitola and reached the second floor windows. (Photo courtesy of the Capitola History Museum).

The Swanton Beach Park (today Natural Bridges State Beach) recently completed seawall was badly damaged by the pounding surf. On Main Beach, two bandstands, the seawall and railing, two restaurants, and picnic tables and benches were carried out to sea by the high tides and surf. The Esplanade on the beach, the steps bordering the casino and boardwalk were washed away (*Figure 4.34*). The extreme high tides and storm seas backed up the San Lorenzo River so high that Bixby Street, the lower end of Ocean Street, and Pearl and Jessie streets were under one to two feet of water.

FEBRUARY 4, 1931

Waves loosened pilings on the Santa Cruz Municipal Wharf and damaged the front of the Casino.

FIGURE 4.34. The steps to the casino and boardwalk were washed away by the large waves and high tides of the February 1926 storms. (Photo courtesy of the Santa Cruz Beach Boardwalk Archives).

While damage was widespread around the northern bay because of the approach of the storm and waves from the west, the north side of the Monterey Peninsula was sheltered. No damage was recorded in the harbor even though the storm waves battering the coast were some of the largest ever observed. On the Carmel side of the peninsula it was a different story with Carmel Beach being completely underwater. A pier was also damaged at Moss Landing.

FEBRUARY 14-16, 1927

At the time this was reported to be the most violent storm in the history of the Pacific coast. During high tide, breakers rolled to the Esplanade and dashed against the Casino at the Boardwalk. The concrete seawall at Seacliff Beach was destroyed.

DECEMBER 9-10 & 23-29, 1931

A violent storm affected much of the California coast. Waves reached 20 feet above mean lower low water producing considerable damage. Low areas along East Cliff Drive were flooded and damaged with sections lost. In Capitola, damage was confined to the waterfront area as "*waves smashed in several doors and windows in the first row of the Venetian Court apartments, drenching the furnishings and compelling a hasty exodus of residents there*". The casino suffered considerable damage. A number of pilings were also broken loose from the Capitola Wharf, which led to its being posted as unsafe for vehicular traffic. The Esplanade was blocked to vehicles because of the enormous quantity of driftwood strewn about. A large part of the new timber bulkhead at Seacliff, which had been built earlier in the year, was destroyed and the concession building and bathing pavilion wrecked. The concrete ship *Palo Alto* was shaken loose and moved about three feet as if "*impelled by the spirit of the sea to fulfill its destiny and start moving*". Cottages and concessions at New Brighton State Beach were damaged.

On the 24th of December high winds were reported to have beached several boats inside the

Monterey Harbor. More boats were reported capsized and sunk at their moorings on the 25th. In the following days 200 feet of a loading pier at a cannery was torn out; the back of the Ocean View Hotel suffered damage. The boardwalk and outer end of Del Monte Bathhouse pier were destroyed. 17-Mile Drive near Fan Shell Beach was badly torn up by waves and littered with boulders and surf-driven logs battered down the door of the Country Club Bathhouse in the same area. Ocean View Boulevard near Point Piños was impassable because of logs and boulders, and the railroad tracks were undermined in Pacific Grove in several places.

FEBRUARY 10, 1938

A storm with winds of up to 70 miles per hour whipped through Santa Cruz and in four hours uprooted 500 trees, while massive seas pounded the waterfront from Capitola to Aptos.

JANUARY 4, 1939

Capitola seemed to have taken the brunt of this storm as the ocean swept over the Esplanade during the night and brought sand and seawater into the dance floor and up to the bandstand. High tide and waves carried sand and rocks, some six to eight inches in diameter, a block and a half into downtown. Sand and rocks were also washed into the lower terraces of the Venetian Court and onto the covered porches of the Casino on the waterfront. Main Beach in front of the Santa Cruz Beach Boardwalk was littered with logs and debris and the boards covering the supporting pilings were again torn off by wave action (*Figure 4.35*)

JANUARY 8, 1940

The Capitola waterfront was lashed by one of worst storms in recent years this morning leaving a trail of havoc (*Figure 4.36*). As the tide peaked, the Capitola Casino on the Esplanade "*began to collapse as immense waves pounded it unmercifully and by noon it was a complete wreck*" *(Figure 4.37)*. A.V. Woodhouse, president of the Capitola Amusement Company, said plans had already been made for rebuilding the structure for the next season.

In Santa Cruz, waves stripped the boards from the front of the Casino, carried the sturdy Cowell's Beach steps far out to sea, crushed the miniature railway structure near the Municipal Wharf, also doing considerable damage to the wharf. At high tide, the waves swept over the Esplanade, depositing huge quantities of debris and mud as far up as the entrance to the Casa del Rey hotel. Water and sand were washed through a door and into the penny arcade.

DECEMBER 26-27, 1940

The crux of local weather problem was at Seacliff where 80 feet of the approach to the old pier and concrete ship were destroyed as logs up to 10 feet long were tossed onto the oceanfront road. Houses were damaged, and two sections of a new timber bulkhead were ripped out. At Moss Landing houses were under a foot of water. Sections of East Cliff at Schwan Lake collapsed. Timbers along the Boardwalk were broken loose.

JANUARY 8-13, 1941

At Seacliff the beach was eroded to bedrock. Another new timber bulkhead (the 3rd) was completed in August of 1940, but the January storms destroyed about half of the new bulkhead less than a year after it had been built *(See Figure 4.27*).

FEBRUARY 26-28, 1941

Heavy winds and gigantic waves. Breakers smashed the Casino steps. West Cliff Drive was closed due to collapse near the blowhole between Columbia Street and Woodrow Avenue, which appeared to extend under West Cliff Drive.

FIGURE 4.35. Storms in 1939 washed under the boardwalk in Santa Cruz, removing the boards and littering the beach with driftwood. (Photo courtesy of the Santa Cruz Beach Boardwalk Archives).

DECEMBER 24-25, 1942

Damage in Monterey was primarily to boats in the harbor where four purse seiners broke loose from their mooring and were driven aground, with two being total losses. A number of other boats were damaged as they were knocked against one another.

FIGURE 4.36. Waves in a January 1940 storm washed onto the Esplanade in Capitola and left major structural damage to the Casino. (Photo courtesy of the Capitola History Museum).

FIGURE 4.37. A different view of waves damaging the Capitola Casino in January 1940. (Photo courtesy of the Capitola History Museum).

DECEMBER 8-9, 1943

Very strong northeast winds resulted in Monterey Harbor damage estimated at $14 million (in 2017 dollars). Forty boats were damaged or lost, including three large purse seiners. Many pilings on the Monterey Municipal Wharf were damaged and planks were broken loose from Fisherman's Wharf.

JANUARY 28, 1947

Northerly gale force winds led to capsize and beaching of a 43-foot fishing boat. An 80-foot long section of dike holding back harbor dredge spoils washed out in Monterey.

AUGUST 1, 1949

The Sentinel reported the "*heaviest summer ground swells in more than 20 years*" when 18-foot waves tossed surfers around as the municipal lifeguard corps had its busiest weekend in history. One swimmer drowned and a number of others were rescued.

OCTOBER 26-29, 1950

Northerly gale winds accompanied by giant waves pounded the Monterey Peninsula. In Carmel, waves crossed Scenic Drive, which is about 25 feet above sea level, and also crossed Ocean View Boulevard in Pacific Grove.

Waves 10 to 15 feet high caused considerable shoreline erosion and swept across Aptos Beach Drive at Rio del Mar, carrying fence posts across the drive smashing against the beachfront homes. Some windows were broken and at least one home had seawater in the living room. Some of the supports for the Seaside Company's bandstand were washed away. The Pleasure Pier was damaged, as was the Municipal Wharf, due to the combination of high tides and huge waves. Twin Lakes Beach also felt the impact of the storms as waves washed over East Cliff Drive.

FEBRUARY 23, 1953

Damage was concentrated in Monterey where seven large fishing boats broke loose from their moorings and were blown into one another. High winds and rough water then drove four of the largest boats onto the beach with significant losses ($4.5 million in 2017 dollars).

NOVEMBER 13, 1953

Southerly winds. Pleasure Pier at Santa Cruz was damaged and waves overtopped the seawall at Capitola and eroded the beaches.

FIGURE 4.38. Storms during the 1958 winter washed under the Ideal Fish Restaurant at the edge of the wharf and damaged some of the supporting pilings. (Photo courtesy of the Santa Cruz Beach Boardwalk Archives).

APRIL 3, 1958

Headlines: "*Spectacular Waves – Damage is Heavy*." High ocean swells caused damage to the entire northern Monterey Bay coast from Santa Cruz to Rio del Mar. In Capitola, waves surging under the pile-supported businesses along the Esplanade "*exploded with hydraulic force, buckling floors, flooding and breaking windows*." At one point in the storm, the waves picked up the Cove Bar from its pilings and then dropped it with a jolt. Waves washed over the low seawall on Capitola Beach and flooded six to eight of the Venetian Court apartments on the beach. The waves also carried mammoth redwood logs over the low wall fronting the Court, which bashed in doors and broke windows.

The Ideal Fish Restaurant at the entrance to the Municipal Wharf in Santa Cruz was hit by large waves that broke windows and weakened the building's underpinnings (*Figure 4.38*). At the Santa Cruz Boardwalk, waves washed up in front of the plunge and Casino buildings and entered the concessions (*Figures 4.39 and 4.40*). They also toppled the front of the bandstand. Breakers washed over East Cliff Drive at Twin Lakes and also at the other low points along the drive.

FEBRUARY 8-10, 1960

Headlines: "*Capitola and Rio del Mar Hit Hard by Gale Winds and Heavy Seas*." Along the Capitola Esplanade, huge waves smashed the beach restaurants and amusement concessions and destroyed the picturesque merry-go-round. Waves rammed against the beach level Venetian Courts, causing extensive damage to at least 12 of the apartments. Once windows were broken, rocks, logs, sand and seawater poured in. A sign was ripped off the end of the wharf, rolled into a ball and deposited in one of the apartments. Cliffs also failed below Grand Avenue along Depot Hill.

Twenty-five luxury beach homes along Beach Drive in Rio del Mar were damaged as the force of the gigantic waves battered garage doors, plate glass windows, and even heavy front doors. Although most of the vacation homes were not occupied, sand and rocks were inches deep in the homes, furniture was shoved against the walls, and one owner returned to find a huge log in his hallway. At Seacliff, the beach, parking area and several trailers owned by vacationing families were damaged.

Five feet of sand was eroded from Cowell's and Main Beach and waves washed across the Boardwalk and Esplanade and dumped water between the Big Dipper and the Wild Mouse rides. Beach Street behind the boardwalk was covered with seaweed carried in by the waves, which crossed the street to reach the motels. Pilings were also damaged beneath the Pleasure Pier. The Castle dining room (on Seabright Beach) was damaged and the beach was completely washed away.

FIGURE 4.39. The 1958 storm waves washed up and onto the steps of the Santa Cruz Boardwalk. (Photo courtesy of the Santa Cruz Beach Boardwalk Archives).

At Davenport, the welded steel Pacific Cement and Aggregates Pier was partially destroyed by the waves, although it hadn't been used in the past decade.

FIGURE 4.40. Waves washing onto the Colonnade along the Boardwalk during the 1958 winter. (Photo courtesy of the Santa Cruz Beach Boardwalk Archives).

On the Monterey Peninsula, high tides and 40-foot waves damaged homes, tore up roads, and washed away the Stillwater Cove Pier in Pebble Beach. Large boulders were strewn along the 17-Mile Drive and portions of the golf courses were flooded with seawater. The seawall near Bird Rock was reported demolished. Numerous small craft were swamped, capsized and beached in the harbor, while Fisherman's Wharf was covered with wave-tossed debris and closed. Ocean water smashed through a 3rd floor window and damaged a room at the Ocean View Hotel on Cannery Row.

FEBRUARY 1963

The west jetty of the Santa Cruz Small Craft Harbor, which was under construction, was damaged as waves broke over it and also destroyed the access road to the jetty. Railroad Wharf at Cowell's Beach destroyed. East Cliff Drive suffered heavy cliff erosion.

FEBRUARY 1969

Large waves from the southwest attacked the new Pajaro Dunes development, with severe erosion at 12 lots. Old automobile bodies were brought in and placed at the toe of the eroding dune for protection.

JANUARY 9, 1978

Headline: "*Savage Seas Attack Santa Cruz Area Coastline.*" This was the first major El Niño winter in decades and the Monterey Bay area coast was attacked by severe storms from the southwest. These had greatest effects on west and southwest facing beaches, and flooded houses and businesses between Capitola and Rio del Mar. Storm waves reached 15 feet in height, and coupled with the year's highest tides, washed ocean water well into inhabited areas. Waves washed across East Cliff Drive at the low points, and the road was impassable at Corcoran Lagoon where a new bridge was under construction. Capitola, Rio del Mar and Pot Belly Beach were the hardest hit. Waves and logs also broke up some of the asphalt parking area directly behind the seawall. Capitola Wharf was damaged as pilings were broken loose. The end of the wharf was torn off, and a shed was teetering precariously over the water. Waves overtopped the low wall fronting the Venetian Court apartments on Capitola Beach and carried sand into the first floors. The seawall and homes along Beach Drive in Rio del Mar were damaged and flooded as fences and decks were washed away when the water swept across the Esplanade. Extensive beach scour undermined the foundation of one beach level home at Pot Belly Beach causing it to subside, and other homes suffered loss of patios, decks, stairways and windows *(See Figure 4.26)*. Several pilings were broken loose from the Santa Cruz Municipal Wharf and the waves broke a central water pipe, which shut off water to restaurants and shops. Waves washed right up to the steps to the Boardwalk, but stopped there.

The timber seawall at Seacliff was overtopped by logs and debris and extensively damaged. Major wave erosion at Pajaro Dunes threatened homes built on the frontal dune at three different locations. Steel barrels, large concrete blocks and sand bags were brought in to protect the homes from further dune erosion. In Monterey, the surf broke windows of some Cannery Row restaurants.

OCTOBER 2, 1979

"*Capitola Coastline Ripped by Huge Waves.*" Freak 12-20 foot waves pounded the Capitola coastline and destroyed or damaged at least eight sailboats as they were torn from their moorings off the Capitola Wharf. Several waves broke over the timber seawall at Seacliff State Beach.

FEBRUARY 13-23, 1980

Six severe storms struck northern and southern California within 10 days. Giant waves lashed the coast in Santa Cruz, completely destroying the parking lot at Twin Lakes State Beach and flooding the village at Capitola. The Capitola Wharf lost ten more feet at its outer end. Seacliff State Beach suffered $3.3 million in damage (2017 dollars) as the entire lower beach portion of the park was destroyed, taking roads, parking lots, and 200 feet of a new seawall. The esplanade area of Rio del Mar was pounded by heavy surf and was closed to non-residents when logs and other debris were washed onto the parking area during high tide. Minor wave damage was reported at the Municipal Wharf in Santa Cruz.

JANUARY 3-5, 1982

Major storms struck the central coast and Monterey Bay. There was extensive erosion from gigantic waves. Most of the damage from this storm was from the prolonged and high intensity rainfall, which included flooding and debris flows in the Santa Cruz Mountains. In Capitola, due in large part to the high flows from Soquel Creek, several of the Venetian Court apartments were damaged and a portion of the seawall collapsed.

JANUARY – MARCH, 1983

A number of severe storm waves coincident with high tides and the largest El Niño in half a century (which elevated sea levels about 12 inches or more for several months along the central coast) brought $250 million (2017 dollars) in damage to the California coast during the first three months of 1983. Santa Cruz County damage was estimated at $32 million.

At Seacliff State Beach, after a number of historic winter storms repeatedly destroyed major sections of the timber bulkhead/seawall, the 2,670-foot long wooden structure was rebuilt again in late 1982 at a cost of $4.4 million (2017 dollars). The new wall was planned to last 20 years, but in late January 1983, within two months of completion, waves, high tides and large logs battered down 700 feet of the new structure *(See Figure 4.6)*. Eleven RV sites were destroyed, restrooms were heavily damaged, and logs and debris were washed back to the old seacliff cutting off access to Las Olas Drive homes. Losses were estimated at $1,900,000 (2017 dollars).

At the south end of Beach Drive in Rio del Mar, the State Beach parking lot was heavily damaged because the timber and piling wall (identical to that at Seacliff State Beach) was destroyed. Logs and waves overtopped the wall, battered the timbers loose and removed fill leading to collapse of the asphalt (*Figure 4.41*). Twenty-six gated homes south of the parking lot were protected by a variety of seawalls that were progressively battered or undermined and failed. Virtually every protective structure was damaged or destroyed. Two houses with shallow foundations were total losses *(See Figures 4.4, 4.5 and 4.7)*, while others lost pilings, windows, decks and stairways. The sewer main beneath the beach serving the beach level homes was exposed and severed, followed by two weeks of raw sewage release. Along Via Gaviota, the back beach development at Aptos Seascape, large waves overtopped the riprap

FIGURE 4.41. Repeated battering by large waves at times of high tides during the major El Niño of 1983 led to partial destruction of the timber bulkhead supporting a parking area at the south end of Beach Drive, Rio Del Mar.

fronting 21 homes. Nineteen of the homes suffered damage, as waves broke through windows, sliding glass door and house fronts (*Figure 4.42*).

FIGURE 4.42. The 1983 El Niño, the most severe in decades, took its toll along the Monterey Bay coastline as a result of elevated sea levels and the simultaneous occurrence of high tides and large waves. These homes built on the beach at Aptos Seascape suffered serious damage as waves used the revetment like a ramp.

Major dune erosion threatened homes at Pajaro Dunes. Erosion took place very quickly so that continued wave action threatened at least 17 homes and 25 condominiums *(See Figure 4.15*). Emergency rip-rap was brought in and placed in front of at least 60 homes (*Figure 4.43*). By late January, up to 40 feet of dune was removed leaving a vertical scarp up to 18 feet high adjacent to the foundations.

Extensive flooding and damage also occurred along restaurants of the Esplanade in Capitola as waves overtopped the low seawall and carried sand and debris well into downtown (*Figure 4.44*).

FIGURE 4.43. Emergency rip-rap was brought in during the 1983 winter to protect the homes at Pajaro Dunes that had been built on the frontal dune.

JANUARY & FEBRUARY 1998

Another powerful El Niño winter storm smashed into the Monterey Bay coastline. Large waves and high tides sent seawater over East Cliff Drive at the familiar spots, Schwan Lake, Corcoran Lagoon and Moran Lake (*Figure 4.45)*. A city bus was hit broadside as it crossed the low spot at Moran Lake, which soaked the passengers with open windows and left water inside the bus. Beaches along West Cliff were scoured down to

FIGURE 4.44. Flooding of downtown Capitola reached a block inland in early 1983 as high tides and storm waves overtopped the seawall.

FIGURE 4.45. Waves and high tides brought seawater over East Cliff in a number of low-lying areas, Twin Lakes in this case. (Photo: David Revell © 2008).

bedrock. Waves up to 25 feet high led to the closure of the Santa Cruz Municipal Wharf as several pilings and some timbers broke loose. Wave attack undermined portions of the bluff along West Cliff removing part of the bike path, and threatening the roadway.

SOME FINAL THOUGHTS ON COASTAL STORM DAMAGE AND SHORELINE EROSION

It is painfully evident from the history of storm damage described above that there are a number of ocean-front areas that have suffered from regular or repeated flooding or wave impact. For the coast from Santa Cruz to Rio del Mar, many of these damaging events have been associated with storms and waves coming from the west or south-west, often during El Niño years. For the Monterey harbor area, damage has been most common when storms arrive from the north or northwest.

For the northern bay, the Santa Cruz Boardwalk and its various facilities – some that have come and gone over the past century – as well as the wharves and piers that have graced the waterfront, have frequently been impacted by high tides and large waves. The low-lying areas along East Cliff Drive, in particular, those places where the road crosses lagoons or "*lakes*" nearly at sea level (Twin Lakes, Corcoran Lagoon, and Moran Lake), have regularly been mentioned in the historical accounts as being flooded or closed. Capitola was also built very close to sea level, with many structures virtually on the beach. The Venetian Court, the Esplanade with its historic businesses and concessions including the Casino and bandstands, the old Capitola Hotel, as well as restaurant row of today, have all been repeatedly hit by waves and high tides, and will continue to be in the future. Sea-level rise is only going to lead to more frequent flooding.

From New Brighton Beach to Rio del Mar, the back beach area began to be developed with small vacation cottages in the 1930s. Development continued over the years with cottages becoming the larger beach homes of today. But these developments, Pot Belly Beach, Las Olas Drive, Beach Drive and Via Gaviota, have all encroached onto the shoreline or the beach. They have been regularly threatened or damaged, and in some cases, destroyed by the combined impacts of high tides, large waves and elevated sea levels during El Niños. Most of these developments are now armored – at least temporarily – by revetments or seawalls, but they took some serious beatings in the past.

Seacliff State Beach sits in the middle of these shoreline subdivisions, on the sand and nearly at sea level, and has taken its share of impacts, only to have the timber bulkhead that supports the picnic and RV area repeatedly destroyed and rebuilt. The sea also has finally broken up the Concrete Ship after about 88 years of constant exposure to all that the ocean could throw at it. The Pajaro Dunes development and its predecessors, Port Watsonville and Port Rogers, also have had a history of storm damage to wharves and development on the dunes.

The shoreline of southern Monterey Bay has been much less intensively developed so that storm damage and coastal erosion have had far fewer impacts. In part this may be related to the existence of the army based at Fort Ord, which owned much of the land. Stilwell Hall, however, was constantly under siege as the bluff edge moved landward at an average rate of about six feet per year, until this historic structure was finally demolished in 2003.

The Monterey Beach Hotel was built directly on the backshore and the Ocean Harbor House condominiums were built on the frontal dune, both sites that are undergoing active retreat. They have both been impacted by large waves during El Niño events since their construction in the 1960s. The Monterey Harbor and Wharf, as well as boats anchored in the harbor, have been repeatedly damaged by northeast and northwest winds. Overall the Monterey Peninsula has suffered very little retreat or damage over the years because of its underlying solid granite foundation.

The coastal storm history of Monterey Bay over the past 150 years has been vividly documented in local newspapers after each severe winter storm. With time, our own personal images and memories fade slowly, but with sea level rising and more intensive development of the oceanfront, the writing is on the wall. Protective structures – such as seawalls and rock revetments – will help, at least for the immediate future, but they are very expensive and will eventually be overtopped, undermined, outflanked or simply fail. We have now armored over 11 miles – or about 25% – of the entire coastline of Santa Cruz County with some type of protective structure. At present day costs of ~$2,500 to $10,000 per front foot (~$13 to $52 million/mile), this is an enormous investment. Whether individuals desiring to protect their own oceanfront homes, and/or the general public who pays indirectly for most of the infrastructure protection, as well as the costs of cleanup and repair, we cannot afford to forget the lessons of coastal storm history and damage around the shoreline of the bay and make the same mistakes repeatedly. All protection stops somewhere and California coastal cities and counties are now beginning to develop response or adaptation plans for future sea-level rise. These will include documenting the most vulnerable areas where relocation or planned retreat will be the only option in the not too distant future.

Chapter 5

Rainfall and Flooding

It's a hard rain's a-gonna fall...
– Bob Dylan

INTRODUCTION

Large damaging earthquakes occur only a few times a century in the Monterey Bay region, but rivers tend to overflow their banks with unfortunate regularity. It is because of this frequency and the large areas affected that flooding has historically been the natural hazard responsible for the most property damage in both Santa Cruz and Monterey Counties. The reasons for this regular and damaging inundation are not a big mystery. Most importantly, we live in an area of seasonally high to very high rainfall; at least historically it rained a lot. The Santa Cruz Mountains, Bonny Doon, Boulder Creek and the upper San Lorenzo Valley typically receive about 50 inches of rainfall annually, with over 100 inches falling in some very wet years; and all that water has to go somewhere.

Fifty to one-hundred inches of rain in a year is a lot of water to deal with, although this pales in comparison with what has been recorded in the area generally accepted as the wettest place on the planet – Cherrapunji, India – right at the foot of the highest mountains on Earth, the Himalayas. Topography plays a big role in rainfall intensity and monsoon storms in south Asia come off the Indian Ocean laden with moisture, hit the Himalayan foothills and start pouring out water. This is a place where almost nobody ever goes outside without a heavy-duty umbrella. It rains so much here that they measure it in feet instead of inches and Cherrapunji holds the world record for one year with 86.8 feet of rainfall! That's almost three inches every day, 365 days a year. They also are at the top of the chart for one month of precipitation with 366 inches, or 30.5 feet. That's a foot every day, all month long. Droughts are not a concern in Cherrapunji. The 24-hour record, however, goes to the steep volcanic island, Reunion, in the tropical waters of the Indian Ocean, with an astonishing 73 inches, or just over six feet in a single day.

At the other precipitation extreme, parts of the Atacama high desert in Chile, sandwiched between the Chilean Coast Range and the Andes have never recorded any rainfall in historic time. They don't sell many umbrellas there.

SOME PERSPECTIVES ON MONTEREY BAY AREA WATERSHEDS AND FLOODING

Not only do the steep watersheds draining into northern Monterey Bay receive large amounts of rainfall periodically, many of the communities or cities were built all, or at least in part, on the natural floodplains of the rivers and creeks: downtown Santa Cruz, parts of Felton, Soquel, Capitola, Aptos, and Watsonville, being the main examples in Santa Cruz County. Moving southwards into Monterey County, the watersheds of the Salinas and Pajaro rivers are much larger but not as steep as those of Santa Cruz County. Although both of these rivers overflow, most of the towns around the bay, with the exception of Pajaro, right across the river from Watsonville, are either at a high enough elevation or far enough away from the rivers that main street or city center inundation hasn't been a frequent occurrence. Flooding takes place along both rivers, but the areas affected tend to be agricultural fields, which thrive on the rich floodplain soils.

It isn't an act of God or rocket science why these streams overtop their banks. It's really not much different than the sink or bathtub overflowing when you forget to turn off the water or when the faucet is putting more water in than the drain can handle. Overbank flow is how all streams deal with the runoff from sustained rainfall that can no longer be contained in the stream channel. Almost no stream or river can carve and maintain a channel large enough to carry all floods; big floods just don't occur often enough to create and sustain a very large channel. Instead, a creek or river will erode a channel which will contain a certain flow of water, usually about the maximum which is carried every year or two. Any flow greater than this will overflow the channel where the water will then slow down, drop out sediment, and over time, build a fertile floodplain. These flat areas adjacent to the large streams form the world's most productive farmland. The Salinas Valley is the nation's salad bowl because of the periodic overflow of the Salinas River over thousands of years, which has been accompanied by the deposition of a fresh layer of fertile sand and silt. These rich floodplain lands in Monterey County's Salinas River Valley provide 61% of the nation's leaf lettuce, 57% of its celery, 56% of its head lettuce, 48% of its broccoli, 38% of its spinach, and 30% of its cauliflower. Without these regular replenishing, although sometimes damaging river floods, we wouldn't have the fertile soils, the crops and this huge economic engine.

THE DEVELOPMENT ON THE FLOODPLAINS OF SANTA CRUZ COUNTY

The Native Americans in the Monterey Bay region observed and understood these flood patterns and landforms and consciously chose not to live permanently along any of the Monterey Bay area streams. They knew they were going to get wet if they did. It took the new immigrants a while to figure this out, however, and many people and communities never have.

On August 28, 1791, Father Fermín Francisco de Lasuén planted a cross about 500 feet from the San Lorenzo River not far from the present site of San Lorenzo Lumber Company on River Street. For some unknown reason he decided this was where the Santa Cruz Mission was to be built. Perhaps it was because it was just a short walk for water and the surrounding soils looked fertile. A crudely constructed chapel was begun there in September 1791. Historical records indicate that "*two years later the hastily built chapel was collapsed by a flood due to heavy rains*". After considerable correspondence and pleading with the Viceroy in Mexico City, who apparently had little concern or sympathy with the inundation problems Lasuén and his colleagues were experiencing 2,000 miles away, they finally did receive permission to relocate the mission to higher ground. The cornerstone of the second church was laid in February of 1793 on Mission Hill. Major earthquake damage, blamed for some reason on the great 1857 Ft. Tejon shock 225 miles away, combined with general wear and tear on the adobe bricks, gradually led to collapse of most of the original structure. In the 1930s, local money was raised to construct the present two-thirds scale replica based on an old drawing of what the original mission was believed to look like. Lots of other California cities, Carmel for example, were left with beautiful historic missions. Santa Cruz ended up with a scaled down replica of what we think it looked like, and this all started when the original chapel was unfortunately built down on the San Lorenzo River floodplain. They should have just asked the local natives before they started.

Most of the early development of the city of Santa Cruz took place well above the river bottom, on the bluffs on either side of the floodplain and present downtown. The area along the San Lorenzo River bottomlands occupied by the city of Santa Cruz today was avoided for about half a century. But around the time of the California Gold Rush, encroachment of the first brick building (a blacksmith shop built by Elihu Anthony in 1848) took place on upper Pacific Avenue in the area near the present town clock. Once this bold but somewhat foolish step had been taken, the downtown development trend was underway and the process quickly proliferated. The attraction of flat fertile land near the river, easy access to fresh water, and the presence of a few convenient businesses, led to progressively more home and commercial construction on the flats. In 1866, the decision was made to build the new county courthouse on Cooper Street rather than on Mission Hill. Consciously or unconsciously, the die was cast to build the city on the floodplain and directly in the path of the San Lorenzo River, barely a trickle in many years, but a raging torrent in others.

Once this downtown development trend began, the process mushroomed quickly. A few feeble efforts were made to control the path of the river; Bulkhead Street, for example, which runs diagonally one block behind the town clock, was named after a former timber bulkhead built in this area in an attempt to control the course of the river and divert it away from the developing city center *(See Figure 5.1)*. The entire downtown flat, however, from the town clock across the river to where Water Street climbs to Branciforte Drive, from the high school football field along Laurel Street to Ocean Street, has all been formed by San Lorenzo River overflow during countless floods over the past tens of thousands of years. The message from this topography and the flood history that follows is consistent: this area is a floodplain, and from time to time it has been covered with water, sometimes a lot of water.

The flat topography in downtown Soquel looks a lot like Santa Cruz, although on a smaller scale.

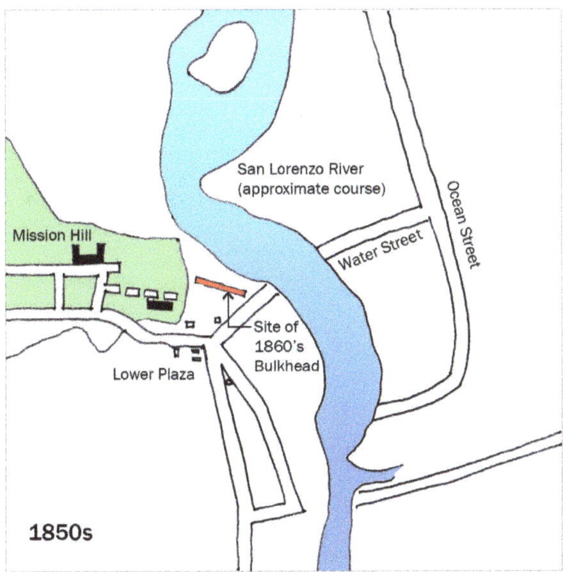

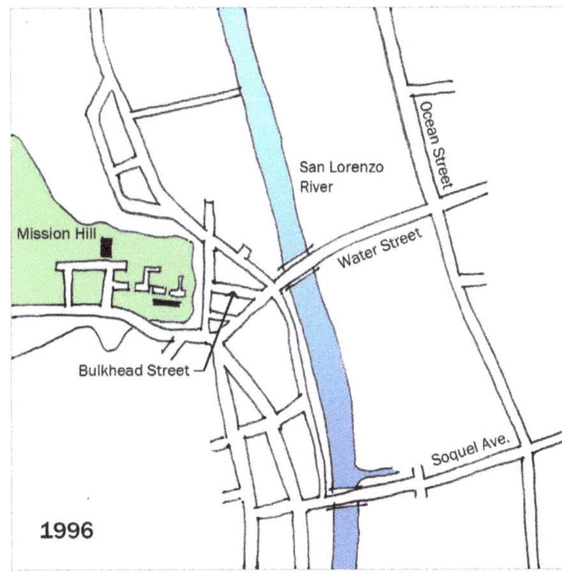

FIGURE 5.1. The historic course of the San Lorenzo River was changed and controlled when the flood control project was constructed. (Illustrations based upon Daniel McMahon's original maps, courtesy of Santa Cruz Public Library System).

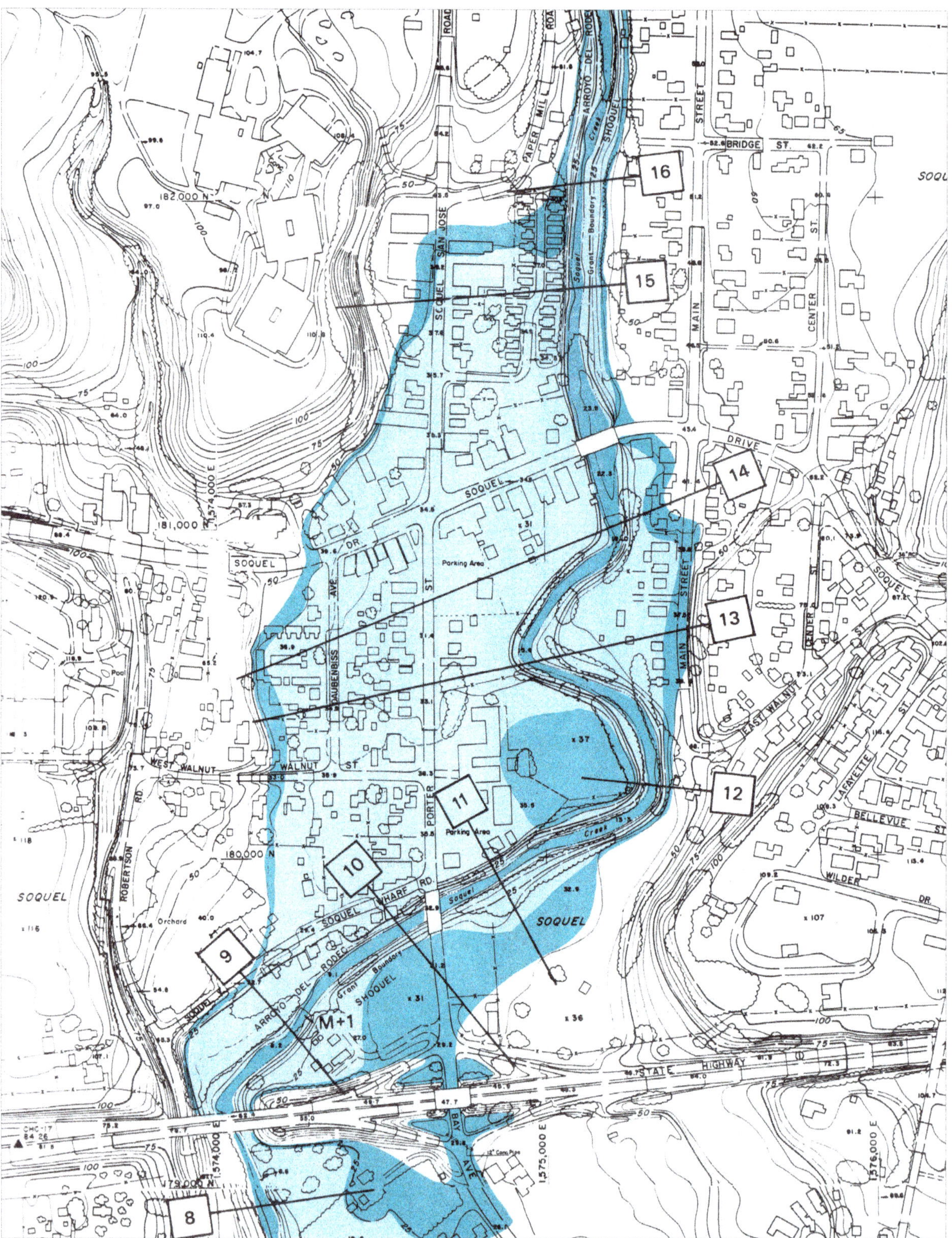

FIGURE 5.2. The U.S. Army Corps of Engineers Floodplain map for Soquel from 1973. The central darker blue line is the actual creek channel; the lighter blue area on either side is their designated 100-year floodplain, and the darker blue bands to either side of that is the 200-year floodplain. (The lines with boxed numbers are cross-sections that are included in the original report).

As you head eastward on Soquel Drive from 41st Avenue, you descend downhill to the business district and as the road levels off, just before crossing Daubenbiss Avenue, you enter the floodplain. Continuing along Soquel Avenue, passing Porter Street and the fire station, you cross over Soquel Creek. After passing Main Street, you start uphill to safe ground and leave the floodplain behind. Virtually the entire business district along Soquel Avenue and Porter Street, from the freeway to the library, and from Daubenbiss nearly to Main Street, has all been repeatedly flooded (*Figure 5.2*). Soquel Creek has meandered back and forth, building this floodplain with sediment eroded from the watershed over thousands of years. It has happened relatively frequently in the past and is guaranteed to happen again. Interestingly, there is an ark of sorts in Soquel, perhaps waiting for the next gigantic flood. On the corner of Soquel Avenue and Main is a church, built by a ship's carpenter in 1870 for $2,700, which internally has some resemblance to a boat.

We could spend pages analyzing why development proceeded as it did; Why Santa Cruz, Soquel, Watsonville, Sacramento or New Orleans, or any of the other roughly 22,000 communities in the United States, which are flood prone, grew up along rivers as they did. In most cases the answers are straightforward: there was water available in the river or stream, whether for domestic or industrial use, for waste disposal, and in the case of larger streams, for transport by boats. The flat fertile land was also usually ideal for agriculture. So the farms, the homes and finally the communities typically grew up in these areas. These benefits don't come without their costs, however.

Unfortunately, very few communities, once established, have ever decided to abandon their floodplain locations. Even after repeated flooding with its destruction and losses, people and businesses don't want to leave. The traditional approach of the Army Corps of Engineers has been to protect flood prone areas through a combination of dams, levees and channel "*improvement*" projects. Trying to subdue nature guarantees full employment for the long term. The damage from the 1993 floods along the Mississippi River, where more money has been spent on projects labeled as "*flood control*" than in any other place on Earth, provided convincing evidence that the flood control approach is not always effective. Unfortunately, we have come to rely on dikes and levees, which typically have had strong political support because they represent federal money and jobs to congressional districts, which have in the past been well received, despite their limitations. A cruel fact of life with floods and any geologic hazard for that matter, is that while we can attempt to provide protection through construction of one project or another, every protection effort and every engineering project has its limits and lifespan. And they all cost money, usually a lot of money; and government at all levels is typically in debt or severely constrained financially.

People living on floodplains need to understand that all flood protection stops somewhere. We simply cannot afford to provide complete protection to all communities from all flood events. The practice has been to provide protection from what is the engineer's best estimate of the 100-year flood, or a flood level that is the worst to expect every 100 years on average. This is not to say that 100-year flood occurs on a regular schedule, systematically every hundred years, but instead that if we had enough years of record (which we rarely do) we could expect to observe floods of this magnitude roughly twice in 200 years, 3 times in 300 years, etc. In reality, we might get two large floods within 25 years and not see another as large for another 150 years. While there are records of large floods in Egypt and China that go back hundreds of years, California's recorded history is a good deal shorter. And we haven't been measuring the size of floods for most of the region's rivers for anywhere close to a hundred years, making it challenging to know how large the 100-year flood might actually be. Complicating the difficulty of determining the 100-year flood is the climate change the planet is undergoing, which will in all likelihood lead to more concentrated winter rainfall and runoff, and larger and more frequent floods.

While we can estimate the size of a hundred-year flood for Soquel Creek and try to build a levee or flood wall to hold it back (if we had the inclination and money), the 110-year flood can come along, surprise everyone, overtop the levee and flood Soquel anyway. We can't afford to build protective structures for all of these events, or for all communities. The flood history of Soquel indicates that it is apt to be flooded every 10 years, not every 100 years. Thus the risk of waking up to floodwater in your mobile home or downtown business is very high, and Soquel has no flood protection – zero. Adjacent to the historic covered bridge in Felton is a small neighborhood known as Felton Grove. It historically has been flooded about every three years on average.

The city of Santa Cruz has been dealing with the shortcomings of an inadequately understood river and a poorly designed flood control project for over fifty years. In the spring of 1954, almost two years before the disastrous 1955 Christmas flood, the Army Corps of Engineers applied to Congress for $2.26 million for the construction of a flood control project on the lower 2.5 miles of the San Lorenzo River and lower Branciforte Creek. A preliminary design had already been completed using flow data from a 1940 flood. The December 1955 flood interrupted design work and necessitated re-evaluation of the flood capacity of the project, but also provided the Corps of Engineers with an even stronger justification for the flood control plan. The funds were appropriated and construction began in 1957 after the flood capacity of the proposed San Lorenzo River channel was increased 25% and that for Branciforte Creek was increased 110%, making it clear that the Corps had vastly underestimated the flood potential of both streams. The 1955 flood was larger than had been expected, in part because the calculation of the 100-year flood had been based on only 20 years of stream flow data.

When engineers plan any new flood control project, they first calculate a "*design flood*", which is the maximum amount of water that a dam, or levees in the case of Santa Cruz, are designed to accommodate. The process of calculating the design flood involves some simple math and then a healthy amount of uncertainty. Overestimating the size of the flood to be expected means a bigger and more costly project; underestimating the flood may mean dam or levee overtopping and potentially disastrous flooding.

While the San Lorenzo River levees were designed by the Army Corps of Engineers to withstand the calculated 100-year flood, those calculations were based on only about 20 years of river discharge measurements. Mark Twain once said "*There are three kinds of lies: lies, damned lies, and statistics.*" Designing levees to handle the largest flood that might occur on average once every 100 years with just 20 years of river flow records, involves statistics and considerable extrapolation.

With a design flood in hand, the next step is to plan a channel that will hold that volume of water, usually given in thousands of cubic feet per second. So the Army Corps, in their infinite wisdom, tried something apparently never before done in the history of flood control, the somewhat innocuously sounding "*below-bed containment*". They excavated about 770,000 cubic yards of sediment (roughly 77,000 dump truck loads) from the existing downtown river channel, and then built levees for 2.5 miles upstream from the ocean to the State Highway 1 Bridge. This design was intended to theoretically allow the river to accommodate the calculated 100-year flood without overtopping the levees. By deepening the channel, rather than widening it, the project's area in theory allowed for passage of the 100-year flood while minimizing the amount of downtown area that would be taken up by the river channel. This allowed for urban renewal and more intense use of the adjacent floodplain, which was met with great joy by the downtown business community. Chinatown, which had been built on stilts out over the marshy area adjacent to the river, was damaged during the 1955 flood. This community was quickly demolished and soon replaced with the banks, restaurants, stores and shops extending from CVS pharmacy to Wells-Fargo Bank between Front and River streets (*Figure 5.3*).

While there was mention of "*channel maintenance*" in the documents that were turned over to the City Council by the Army Corps of Engineers, which included keeping excess sediment from accumulating in the new excavated flood control channel, the magnitude and cost of this problem wasn't fully understood or appreciated by the city. After all, the City Council was made up primarily of local people from the business community, not civil engineers. In subsequent years, the channel, which was dredged from two to eight feet below its natural bottom, filled back up to its previous level, and soon only provided protection from about the 35-year flood, not the 100-year event. This wasn't welcome news. And since the late mid-1970s the city, in combination with the Corps of Engineers, various advisory committees and consultants, worked to develop plans that would increase the flood protection provided and also improve the ap-

FIGURE 5.3. The San Lorenzo River flood control project in downtown Santa Cruz under construction in 1959. View looking upstream showing the footbridge and the Water Street Bridge. (Photo courtesy of the Santa Cruz Economic Development Department).

pearance and functioning of the river as it flowed through Santa Cruz. This is no trivial matter; and while several different designs were developed and approved, it was the 1982 floods that led to reconsideration of the level of flood protection provided by the levees.

While various riverfront renewal projects were being contemplated in an effort to help convert the sterile flood control channel into a more scenic and usable part of downtown, the winter of 1981-82 descended on the central coast. Rainfall started early and continued to saturate the region's watersheds through November and December 1981. The San Lorenzo Valley received about 30 inches of rain in the last two months of the year. Then the skies really opened up and the San Lorenzo River, as well as Soquel Creek and Aptos Creek started to rise. Each of these three Santa Cruz County streams combine steep topography with relatively small or short drainage basins such that downstream flooding historically occurred rapidly with little advance warning. Floodwaters can flow through the entire San Lorenzo Valley watershed from Waterman Gap to the river mouth in less than 12 hours. Thus while floods can be devastating here, they usually don't last very long. This probably matters little to someone once their house has filled with five feet of water; whether their living room and TV have been underwater for four hours or four days, the results are still disastrous. Floods along the Mississippi River system are different beasts. In order for there to be flooding at St. Louis, for example, the entire middle third of the United States, thousands of square miles has to have received billions of gallons of rainwater, and a bathtub this size takes a long time to drain. During the 1973 floods, for example, river levels stood at flood stage and entire communities were under water on the lower Mississippi River for over two months!

HISTORIC FLOODING IN THE MONTEREY BAY REGION

The main flood season for the Monterey Bay area typically extends from November through April when about 90% of the rain usually falls. Historically, January has been the wettest month. For Santa Cruz County, the steep watersheds, occasional high intensity rainfall and short river lengths have the potential to produce downstream flooding in those communities built on the floodplains with relatively little warning. The total length of the San Lorenzo River from the headwaters to the mouth is only about 29 miles, so runoff from rain falling on the highest reaches of the watershed reaches Santa Cruz very quickly.

During the 1955 floods, the discharge of Soquel Creek increased from 2,000 cubic feet/second to 12,000 cubic feet/second in just four hours. For a more understandable reference point, a flow of 12,000 cubic feet/second would fill an Olympic swimming pool in just over 13 seconds. The first indication of the flood of January 3-4, 1982 in the Old Mill Mobile Home Park was when elderly residents noticed water flowing in under their doors and floating their mobile homes off their foundations (*Figure 5.4*).

Another factor somewhat unique to the central coast, which has increased the extent and damage from flooding, has been the wooded character of the Santa Cruz Mountains as well as the history of logging and land clearing. The San Lorenzo River is spanned 16 times by bridges. Each of these crossings represents a potential location for a logjam when the trees, logs, stumps and other debris carried by floodwaters can become trapped in these narrow constrictions. During the 1982 floods, overflow of the levees in downtown Santa Cruz was narrowly averted as cranes operated all night on the evening of January 4th at both the Highway 1 and the Riverside Avenue bridges, keeping large logs and trees from lodging on the upstream side or under these low bridges and then backing up floodwaters (*Figure 5.5*).

FIGURE 5.4. Aerial view of the mobile home park in downtown Soquel following the January 1982 floods. A number of mobile homes were floated off of their foundations and suffered major damage.

FIGURE 5.5. During the January 1982 floods a crane was set up on the Riverside Avenue Bridge to remove trees or logs that could potentially block the flow under the bridge's low arches. This bridge was subsequently replaced. (Photo: Pete Amos and Bill Lovejoy / *Santa Cruz Sentinel* © 1982).

The Soquel Drive Bridge in the center of Soquel has a center pier that created a massive logjam in the winter of 1955, diverting almost the entire flow of the Soquel Creek through the downtown area (*Figure 5.6*). Under the headline "*Soquel Merchants, People Suffer Heavily from Flood*" on December 25, 1955, the *Santa Cruz Sentinel* News reported:

"A tremendous log jam, which developed at the bridge, did its part to divert the muddy floodwaters. Hundreds of tons of debris including giant pine trees, a four-room redwood house and five auto court apartments, which set cockily atop the pile like toy blocks, made the jam. The swirling waters swept in behind a phalanx of business houses on Main Street and roared through with such a powerful flushing force that it destroyed the interior of some 15 stores and offices. The flash flood caught many owners in their establishments. They took to the rooftops and attics and shivered helplessly through the night.

"John Pollock, owner of Soquel Lodge had just remodeled two of his cabins. The water swept

them and five others away from their foundations and onto the giant jam like they were toy boats... For an hour on Friday afternoon, I watched dynamite experts prepare to blow the huge jam apart. The spectacular danger involved in the project curiously drew large crowds.

"When a delay developed, at the scheduled 4 p.m. "blow-up" zero hour, tension began to build up. I drove to Capitola to see the situation there. A general evacuation had been ordered by the police for those living in the low flat lands bordering the river. Evacuation was ordered because it was believed that the Soquel logjam had built a tremendous amount of backwater pressure. They figured that when it was relieved suddenly the water and tons of debris would rush down the mile separating Soquel Creek from Capitola and burst over the banks. The otherwise busy business district was deserted. A few storeowners stood by their front doors waiting and wondering what was to happen.

"But high on the bank along Capitola Road and along the Southern Pacific trestle which crosses the river near the Capitola bridge, were lined hundreds of people, standing in the driving rain waiting for the dynamite to blow the log jam. Like spectators to some thrilling paid event, they waited to see the river vent its pent up anger. Then word got around that the authorities decided that

FIGURE 5.6. The Soquel Drive Bridge in downtown Soquel became a dam for logs and debris carried downstream from the watershed during the 1955 floods. The floodwaters were diverted around the bridge, washing out the approach. (Photo courtesy of Carolyn Swift).

FIGURE 5.7. In the January 1982 flood the Soquel Drive Bridge again became a dam for logs and debris, forcing the creek flow through downtown, including flooding the mobile home park (see Figure 5.4).

the blast to release the jam was too risky, and they slopped slowly home."

During the January 1982 floods, another massive logjam formed at the same restricted Soquel Drive Bridge crossing (*Figure 5.7*). Logs and debris filled the entire width of the creek channel and extended about 300 feet upstream. While flooding of downtown Soquel had already taken place, the logjam diverted the main flow of the creek to the west and through the Old Mill Mobile Home Park where dozens of coaches were floated off their foundations, and twisted around while suffering major water damage *(See Figure 5.4)*. The floodwater deposited two to three feet of sand and silt in the downtown business district as floodwaters reached depths of over four feet (*Figure 5.8*).

FIGURE 5.8. Soquel Drive at Porter Street on January 4, 1982 following the overflow of Soquel Creek. The Soquel Drive Bridge (see Figure 5.7) is visible in the background.

Hydraulic studies of the flood control project on the lower San Lorenzo have shown how the low arches of the Riverside Avenue Bridge and also the upstream portion of the Water Street Bridge could intensify flooding by restricting flow and backing up floodwaters. This happened at both bridges in early January 1982. The Riverside Avenue Bridge was ultimately demolished due to structural damage suffered during the 1989 Loma Prieta earthquake and was subsequently replaced with a bridge having a less constricted opening. The older upstream portion of the Water Street Bridge was also replaced during post-flood improvements of the San Lorenzo River flood control project when levees were also raised.

MAJOR HISTORIC FLOODS IN SANTA CRUZ COUNTY

Most of the accounts of early flooding in the Santa Cruz area are contained in the files of the *Santa Cruz Sentinel* and its predecessors. Since the newspaper did not begin publication until 1856 and local populations were quite small in the early decades of the 1800s, data on pre-1856 floods are scarce. Scattered writing by local historians provides the only information on earlier events. Daniel McMahon and Frank Perry have both researched the flood history of the San Lorenzo River from the local newspapers and have very useful summaries posted on the Santa Cruz Public Library website. The earliest flooding events where any records exist are those of the 1791-93 years when the San Lorenzo River damaged the earliest version of the Santa Cruz Mission that had been built on the flats below Mission Hill.

There are a few descriptions of floods on the lowlands (now downtown Santa Cruz) in the years before 1862, but they are generally quite brief as there was very little development in this area.

Floods did occur with striking regularity, however, that "*covered all the lowlands*" in 1822, 1832, 1842 and again in 1852. Despite this repeated flooding, by the 1840s a few houses had been constructed down on the flats, and in 1849, the first brick building was constructed by Elihu Anthony near the present location of the town clock.

One of the earliest accounts of flooding is from an 1852 flood by a Mr. E.S. Harrison:

"The winter rains had fairly set in; for two weeks there had been a steady downpour; the creeks that a short time previous were dry were now running full, and torrents of water were coming down from the mountains and rushing with great velocity, filling up the various ravines and creeks, rendering them for a time impassable. Work was pretty much suspended, and about all of the population were assembled at the only tavern in the place."

The Record Floods of the Winter of 1862

The winter of 1861-62 proved to be especially severe, not only in the Monterey Bay region, but throughout the entire west from what was to become the state of Washington, to northern Mexico, and from the coast inland to Idaho, Nevada, Utah, Arizona and New Mexico. The combination of weeks of continuous rain and snows in the higher elevations that began in November 1861 and continued into January 1862 was amplified by warm intense storms that melted much of the snow, contributing further to runoff and flooding. Devastating floods occurred throughout California, perhaps the worst in history, with a huge lake (described as being 300 miles long and averaging 20 miles in width and up to 30 feet deep) covering 5,000 to 6,000 square miles of the vast Central Valley.

The new capitol of Sacramento, which had been built at the confluence of the Sacramento and American rivers, was completely under water for sixty days. While a levee had been built along the west side of the city to protect it from flooding, waters entered Sacramento from the higher land on the east, with the levee acting as a dam to keep the water in. A passenger on a riverboat heading up the Sacramento River during the flood reported that:

".. I was a passenger on the old steamer Gem, from Sacramento to Red Bluff. The only way the pilot could tell where the channel of the river was, was by the cottonwood trees on each side of the river. The boat had to stop several times and take men out of the tops of trees and off the roofs of houses. In our trip up the river we met property of every description floating down – dead horses and cattle, sheep, hogs, houses, haystacks, household furniture, and everything imaginable was on its way for the ocean. Arriving at Red Bluff, there was water everywhere as far as the eye could reach, and what few bridges there had been in the country were all swept away."

San Francisco, which has an average annual rainfall of 20 inches, received over 49 inches of rain that particular winter. Locally, floods in early January swept away all dams on the San Lorenzo River as well as numerous barns and large trees. All of the dams on Soquel Creek were lost and water flowed four feet deep through Soquel Village, virtually identical to the levels reached 120 years later in the floods of 1982. Mills, flumes, houses, barns, a school and the town hall were all destroyed. A massive pile of debris was carried to the ocean where it then washed back up at Soquel Landing (the future site of Capitola).

The entire lower section of the Pajaro Valley was underwater and orchards, buildings, dams and mills were destroyed along Corralitos Creek as large numbers of livestock were drowned. Thirty-four years later it was reported that "*old residents remember the winter of 1861-62 as the severest ever known since the settlement of the country by Americans*".

Rain and flooding continued into the latter part of January and the *Santa Cruz Sentinel* at that time reported that: "*The County of Santa Cruz has suffered considerable loss from the devasta-*

tions of the late flood, and were it not for the general and extensive damage sustained throughout the state, the destruction of property here would be considered enormous". By the time of the 1862 flood, the business district had relocated onto the flats, and the juncture of Water, Willow (now Pacific Avenue) and Main (now Front) streets had become "*the lower plaza*" (in contrast to "*the upper plaza*" where the mission and Holy Cross Church are now located). Construction of a new county courthouse (the former Cooper House) on the floodplain in 1866 cemented the shift of the city center from Mission Hill to a far more hazardous location on the San Lorenzo River's floodplain.

The downtown residents were surprised to see debris and buildings from upstream floating down the river in the floods of 1862, including a large barn reportedly drifting towards the ocean completely upright. All roads leading into Santa Cruz were damaged or destroyed, as were the bridges. The San Lorenzo Paper Mill (just above the city near present-day Paradise Park) was washed away in the floods. The course of the San Lorenzo was altered by bank erosion such that it ran several hundred feet nearer to town that it did during the previous flood. While a little difficult to envision now with the river's course more-or-less controlled by rock covered levees and directed very linearly through the city, the river historically meandered back and forth across its entire floodplain, as all rivers do. At one time or another over the millennia, the San Lorenzo River eroded its course to the sea and deposited a floodplain from the town clock on the west, to the bluff you ascend on your way up to Branciforte Avenue on the east. If you are driving on Laurel, the flat area extending from the high school soccer/football field on the east, to where Broadway climbs the hill after crossing Ocean Street, is all part of the river's historic domain. And every few years, until the late 1950s when the levees were built, the river would remind the downtown residents who was in charge.

Following the 1862 flooding of downtown and concerns with future flooding of the new "*lower plaza*" as well as land loss from bank erosion, the bold new city took the first major step to control the flow of the river and built a bulkhead or training wall to divert the river away from its Mission Hill path *(See Figure 5.1)*. The remnants of this effort, now over 150 years ago, still exist in name as Bulkhead Street, a short stretch of diagonal asphalt one block north of the town clock connecting North Pacific Avenue with Water Street.

December 1866 Flood

On December 20, 1866 a strong southwesterly gale brought heavy rains to Santa Cruz. On the following morning the *Sentinel* reported that:

"*The San Lorenzo River was booming full of water and driftwood was filling the channel from bank to bank. At four o'clock in the morning the current was within two inches of running over the bulkhead, north of the foundry, and at the low place where Water Street crosses, the bank was submerged.*"

The Paper Mill dam was again washed away suspending operations there until the next June. Many roads were washed out as well as a bridge at the Powder Mill (near the present-day location of Paradise Park along Highway 9). Flooding caused so much disruption in Santa Cruz that for the first time, the construction of levees was considered. The Santa Cruz newspaper in an article from February 7, 1862 titled "*The Inundation and the Remedy*" stated: "*Doubts have been expressed by some, in reference to the financial scheme or raising means to build a levee. With the protection afforded by the embankment, no such risks need be run: The land now worthless.*"

The rains also led to numerous landslides in the Santa Cruz Mountains, especially along the road from Santa Cruz to Santa Clara, which was rendered impassable. Maintaining a highway through the frequently unstable hillsides of the Santa Cruz Mountains in the winter months has a very long history.

February 1869 Flood

A little over two years later, in February of 1869, the San Lorenzo River again overtopped its banks. Water went over the bulkhead north of the foundry (the foundry was located on what today is North Pacific Avenue up against the cliff just below Mission Hill – *Figure 5.9*) and flowed across many parcels of downtown land carrying debris with it. All houses along the river were inundated, and a sandy island in the river east of Front Street was covered with water up to five feet deep. There were homes on this former island and many of the families living there had to be rescued by boat. Logjams were widespread and one of them destroyed the footbridge at the Riverside crossing to Branciforte. Landslides again were widespread in the mountains on the Santa Cruz to Santa Clara road. In Soquel Village a new bridge was washed away and roads were impassable.

February 1878 Flood

The *Sentinel* of February 23,1878 reported that: *"Monday, the rivers were at flood-tide, but not as high as the previous Thursday. Between then and eleven o'clock in the morning, the tide full, rolled in across the Santa Cruz Railroad doing considerable damage. On Tuesday, Pacific Avenue was flooded and, "any man that has attempted to wade through... at any time during the past forty days, is lost to society, lost!"*

FIGURE 5.9. The first foundry in Santa Cruz located on North Pacific Avenue below Mission Hill (1879). (Photo courtesy of the Romanzo E. Wood Collection, Meriam Library, California State University Chico).

January 25, 1890 Flood

The winter of 1889-90 produced a remarkable 63.2 inches of rainfall in Santa Cruz and 120 inches in Boulder Creek. Because of the way in which the rainfall was distributed throughout the winter months, however, there was only one flood, on January 25. Older residents compared it to the 1862 flood in magnitude. Nearly all of downtown Santa Cruz was underwater; many bridges were destroyed and the city was completely cut off from railroad communication. People were stationed on bridges to guard against logjams by dislodging logs and debris. Soquel Creek and the Pajaro River also flooded on this date, as was the Esplanade area of Capitola (*Figure 5.10*). Damage consisted of washed out bridges and roads and flooded merchandise in the downtown sections of Soquel and Watsonville.

FIGURE 5.10. The 1890 Soquel Creek flood in Capitola, looking upstream from the beach with the railroad trestle in the background. (Photo courtesy of the Daubenbis Collection, Capitola Historical Museum).

March 23, 1899 Flood

The reporting of disasters became quite poetic during this flood as seen in the March 24, 1899 edition of the Sentinel:

"The booming San Lorenzo River and Branciforte Creek were on a wilder rampage Thursday than they had displayed in years. They had indulged in a sort of turbulent gaiety, which would not be restrained. From quiet, placid streams they developed into bold, aggressive, defiant rivers. They had become expansionists with a vengeance. From a quiet sleep they had awakened to become wild, angry, roaring, reckless, impetuous masses of water, which bulkheads and sandbags could not restrain. They leaped over their banks with the agility of acrobats, and in their caresses embraced orchards, gardens, fences and houses."

The water apparently did not reach Pacific Avenue because the land between the avenue and the river had now been filled in. Lower portions of downtown were underwater, however, and the railroad bridge was slightly displaced by a large logjam. A young boy reportedly fell from this bridge and drowned during the flood.

March 7 1911, Flood

This flood occurred relatively late in the winter and saw both the San Lorenzo River and Branciforte Creek top their banks after a storm the night before. The San Lorenzo River rose at a fast clip at 7.5 inches per hour and "*overflowed its banks all along the line*". Water overflowed onto lower Pacific Avenue, flowed down Spruce Street (immediately south of Laurel between Cedar and Front streets) and flooded all of the flats bordering the river. Most of the damage consisted of lost land along the river, damaged bridges and downed communication lines. The latter led to the isolation of Santa Cruz from the rest of the world for three days as mail, train and telegraph service were interrupted.

Residents along Branciforte Creek claimed that the floodwaters hadn't been this high in years. Water overtopped the Berkeley Way Bridge and flowed down Market Street to River Street, flooding "*the whole flat*" and forcing people to evacuate. As is often the case with the streams draining the steep Santa Cruz Mountains, there was little advance warning of the floodwaters that were to come. One resident reported that the water was high but not dangerously so at 11:00 p.m., but by 11:15, the water had overflowed the banks and surrounded their barn forcing them to move their horses to higher ground.

In Watsonville, the Pajaro River was the highest in memory in the words of some residents. Overflow extended as far as three miles in every direction. Railroads and bridges were washed out and the Pajaro Valley was reported to have again resembled an inland sea (*Figure 5.11*)

FIGURE 5.11. Heavy rains in early March 1911 led to the Pajaro River overflowing its banks and flooding downtown Watsonville all the way to the main plaza. (Photo courtesy of the Capitola Historical Museum).

January 1914 Floods

Santa Cruz County was pelted by storms during nearly the entire month of January. Although the San Lorenzo River didn't flood Pacific Avenue, water damage was still extensive. On January 1, Soquel Creek again overflowed its banks and flooded residential areas along the stream. The same was true for Branciforte Creek. A high tide coincided with peak flow in the San Lorenzo leading to the inundation of the lower sections of the city.

The second flood of the month took place in Watsonville on the 18th of January, and the area devastated was even more extensive than that area flooded in 1911. Water was six feet deep in some parts of town and the entire Pajaro Valley from Watsonville to the beach was flooded. Continued rain kept water levels high and the floodwaters didn't begin to recede until January 27.

December 1931 Floods

On Christmas Eve, 6.6 inches of rain fell in Ben Lomond in 24 hours. In Soquel a large tree fell across Soquel Creek and sent water over the banks. Three days later, another storm forced Soquel Creek over its banks. The floodwaters in downtown Soquel were the highest in 20 years and covered a large area from the Soquel Drive Bridge to the library on Porter Avenue. Wooden cottages at the Willowbrook Villa camp were pushed off their foundations by floodwaters, and one of the cottages floated downstream lodging against the bridge and damming up considerable debris. This obstruction was later dynamited to keep the water flowing. All of the stores in Capitola were flooded and the Italian Gardens near Paradise Park were damaged when the San Lorenzo topped its banks. The area at the intersection of Riverside Avenue and Ocean Street was underwater as the river filled the floodplain (*Figure 5.12*).

February 13, 1937 Flood

Soquel Creek overflowed its banks again due

to a logjam at the Soquel Drive Bridge and reached its highest level since 1931. The downtown sections of both Soquel and Capitola were flooded although damage was reduced by a massive sandbagging effort.

FIGURE 5.12. View from Ocean View Avenue of the San Lorenzo River during the flood of December 1931 at Barson Street Flats and across the river towards Beach Hill. (Photo courtesy of the McPherson Art and History Museum).

December 10, 1937 Flood

Before the year was out the heaviest 12-hour rainfall in Santa Cruz's recorded history brought 3.35 inches to the city, and within 36 hours, 8.46 inches had fallen. Ben Lomond had 9 inches of rain in 24 hours, and both the San Lorenzo River and Soquel Creek went over their banks again. The San Lorenzo flooded downstream from the Riverside Avenue Bridge, probably because of the high tide conditions. A cabin was washed into Soquel Creek and destroyed. Five bridges were washed out in the south county.

February 1938 Flood

"An ugly, turgid San Lorenzo River was receding last night after reaching flood proportions that broke all records for 15 years. Swelled by heavy mountain rainfall of nearly 20 hours, the runaway stream picked up huge logs and assorted debris to batter them against none-too-secure bridges from Felton to the ocean at Santa Cruz."

Crews of city employees worked under floodlights and, with the aid of a tractor, continued to dislodge logs and debris, which were piling up against the upstream side of the Ocean Street Bridge. Loss of some of the pilings that had been torn out by the raging waters led to sagging and partial collapse of the bridge (*Figure 5.13*) as well as flooding along East Cliff in the Barson Street Flats area (*Figure 5.14*).

"In Felton, Mrs. H.A. Witham, operator of the Felton Grove resort, said her loss was quite heavy and that 13 cabins were full of mud and water and would be expensively renovated. There were 20 homes and cottages in that area that had mud and silt on the floors but none were moved from their foundations. The San Lorenzo River swept over the Sycamore Grove auto camp, washing away tables and benches, and continuing over the grounds of the Riverbank auto camp at Water Street. Mrs. S. M. Little reported that her office was flooded and that the river was up to the cabins. At Blaine Street the homes of George Barth and other residents were inundated for the first time in 16 years.... considerable damage was done in the beach vicinity where the river had cut a wide channel near the railroad trestle to undermine the boardwalk. About 100 feet of the promenade was hanging in the air."

This may have been one of the earliest accounts of the flooding in Felton Grove, which in subse-

FIGURE 5.13. View along East Cliff towards the Ocean Street Bridge, which has partially collapsed during either the 1938 or 1940 floods. (Photo: Ed Webber, courtesy of Santa Cruz Beach Boardwalk Archives).

FIGURE 5.14. 1938 flood looking over the Barson Street flats area across the San Lorenzo River to Beach Hill and the Boardwalk. Note same house as in Figure 5.12. (Photo: Ed Webber, courtesy of the Santa Cruz Beach Boardwalk Archives).

quent years has been inundated during virtually every significant flood event. Adjacent to the old covered bridge, Felton Grove is one of the lowest neighborhoods along the entire course of the San Lorenzo River and actually lies not on the 100-year floodplain, but about the 3-year floodplain. It has been repeatedly flooded, which has neither deterred the residents from living there, nor the realtors from selling and re-selling the homes and cabins. In recent years some of the homes, but by no means all of them, have been elevated. The floods of 2017 again flooded many of the low-lying homes along the river (*Figures 5.15 and 5.16*).

In Watsonville, the Pajaro River over-flowed and flooded 20 blocks of land filling Main Street with water where rowboats replaced cars (*Figures 5.17 and 5.18*). On the opposite side of the river, the entire town of Pajaro was also under water. On February 12th, an additional 20 blocks were flooded as the main Watsonville levee gave way marooning 1,000 people.

FIGURE 5.16. The debris on the garage door indicates the high water level during the 2017 flooding in Felton Grove. This home had been elevated after the last major flood. (Photo: Nutan Mellegers © 2017).

FIGURE 5.15. The February 2017 flood in Felton Grove. (Photo Nutan Mellegers © 2017).

February 27, 1940 Flood

On February 28, 1940, the *Santa Cruz Sentinel* headlines read "*San Lorenzo on Worst Rampage of Century*". In reading through the flood history from the earlier newspaper accounts, it is painfully evident that both Santa Cruz and Soquel have been regularly flooded, and in many cases, each flood as reported in the newspapers seems to be as large or larger than the previous one. Interestingly, the 1938 flood, just two years earlier had been described above as "*breaking all records for 15 years*".

"One of the greatest river floods in the history of Santa Cruz – and unquestionably the most destructive – smashed its way through the San Lorenzo

FIGURE 5.17. Main Street in Watsonville during the 1938 flood on the Pajaro River. (Photo courtesy of the University of California Santa Cruz Special Collections).

Valley yesterday, bearing logs, branches, dead animals, furniture and debris from ruined summer homes and bridges. With county damage conservatively estimated at $500,000 (equivalent to nearly $9 million in 2017 dollars), *there were at least 100 Santa Cruzans rendered temporarily homeless by the swirling yellow flood.*

"Fed by a cloudburst which brought 11.57 inches of rain to Ben Lo-

FIGURE 5.18. The 1938 flood on Main Street in Watsonville where boats became the only way to get around town. (Photo courtesy of the University of California Santa Cruz Special Collections).

mond in 24 hours, the muddy river first jumped its banks here at 2 a.m. and reached its peak about 6 o'clock last night. Although it was believed the danger point had passed, the Bank of America, Pep Creamery and other Pacific Avenue firms sandbagged their doors against higher waters.

"An all-time record for high water was believed set with a 25-foot rise at Paradise Park.

"The thundering flood swept out 100-foot River Glen bridge at Boulder Creek, then hurtled downstream, taking out the Zayante, Cooper, and Ocean Street bridges on its path to the sea. Simultaneously, slides tore out telephone connections with Boulder Creek and choked the San Lorenzo Drive. Severely hit were Paradise

FIGURE 5.19. The 1940 flood on the San Lorenzo River looking across the lower river from Ocean View Avenue towards Beach Hill – similar view as Figure 5.12. (Photo courtesy of the McPherson Art and History Museum).

FIGURE 5.20. The 1940 flood downtown Santa Cruz near the present county jail.. (Photo courtesy of the University of California Santa Cruz Special Collections).

FIGURE 5.21. The 1940 floods along Ocean Street with the San Lorenzo River and the Boardwalk in the distance. (Photo: Ed Webber, courtesy the Santa Cruz Beach Boardwalk Archives).

FIGURE 5.22. Soquel Creek overflowed and flooded downtown Soquel during the floods of February 1940. View is intersection of Soquel Drive and Porter Street looking east. (Photo courtesy of the University of California Santa Cruz Special Collections).

Park, a number of resorts and auto courts on the outskirts of the city, outer River Street, and the entire Ocean-Barson Street area".

Most of the damage during this flood was to summer cabins along the river from Boulder Creek to Paradise Park. Within the city of Santa Cruz, one hundred people were reported evacuated from their homes in the Garfield Street lowlands, many by motorboats (*Figures 5.19, 5.20 and 5.21*). Prior to the 1955 floods and redevelopment of downtown, Garfield Street ran parallel to Ocean Street and would have run approximately along the front of the County Government Center). Four foot bridges across the San Lorenzo River were destroyed and the Soquel and Water Street Bridges were badly damaged. Six out of nine bridges crossing Zayante Creek were lost. Once again, logs and debris piled against the Soquel Drive Bridge led to flooding of the village of Soquel (*Figure 5.22*).

January 12, 1952 Flood

Minor flooding was reported on this date in the lowest lying sections of the city of Santa Cruz as people were evacuated from their homes on Josephine, Blaine, Garfield, Burnett, and River streets, (Blaine and Burnett streets no longer exist) which historically were usually the first areas to experience San Lorenzo River overflow. The trapping of logs and other debris moving down the river from the San Lorenzo Valley and Soquel Creek at restricted bridge openings has repeatedly led to the overbank diversion of water at these locations. Flooding and damage in 1952 was limited due to the effectiveness of the street department crews in breaking up the logjams at the bridges.

Christmas Flood 1955

The worst flood disaster in the history of Santa Cruz took place in late December 1955 following three days of torrential rainfall that dropped 8.9 inches of rain in the city and 18.3 inches of rain in Boulder Creek. The San Lorenzo River crested at the Felton gauge at 22.55 feet with a maximum flow of 30,400 cubic feet per second (this flow would fill an Olympic size swimming pool in less than 3 seconds; or put another way, if you built a 10-foot high waterproof wall around a football or soccer field, this river flow would fill that up in 14 seconds).

Overflow occurred from the headwaters to the mouth and water levels topped all previous high water marks along Kings, Boulder, Two Bar, Bear and Zayante creeks in the upper San Lorenzo basin. The persistent heavy rains and floodwaters loosened and scoured out riverbank trees and logs, floating them downstream where they again became lodged at channel constrictions such as low bridges. The many logjams diverted the high velocity river flows, undercutting and scouring bridges and road fills as well as homes and other structures. Nearly 300 acres of land in the watershed from Felton upstream was flooded. The total estimated San Lorenzo River basin damage was $8,700,900 ($81 million in 2017 dollars) of which $7,629,600 ($70 million in 2017 dollars) was within the city limits of Santa Cruz.

The reports from the *Santa Cruz Sentinel* on Christmas Day, 1955, provide the most graphic and somewhat literary descriptions of the flood damages:

"The torrenting San Lorenzo River, which spread death and destruction through Santa Cruz dealt severe damage along its path in the San Lorenzo Valley. Nearly 100 homes along the usually placid river were destroyed-some were swept down by the roaring flood. Along the tributaries of the San Lorenzo more homes were torn and splintered. Many families are still isolated with dwindling food supplies. In the dark, murky night of Thursday, the waters rose over the bank and charged with unbelievable force through the thick forest resort parks. It crammed five-room houses against trees, twisted others off their foundations.

"For nearly a day the valley was cut at both ends. Highway 9 from Santa Cruz to Felton is still blocked by trees and great slides of rock and earth. The once beautiful resort area of Sycamore Grove just as you enter the valley from Santa Cruz was a sea of oozing mud and scrambled cabins. The small units were scattered like building blocks. The lower part of Paradise Park, the Masonic tract, was hit and ripped as if the river edge had the teeth of a saw. Three houses disappeared completely...ten houses had their interiors ruined...Another ten will have to be rebuilt...

"Felton Grove...was nearly wiped out. Twelve persons were taken out of the flooded area in boats. Five cabins were carried away, some 20 were destroyed, crushed by the rushing current. The sight of Gold Gulch Park, a subdivision of private homes in Tanglewood (just downstream from Felton) *was startling. Looking down toward the river from the highway, I could see a tumbled muddy mess of expensive mountain homes, shoved here and there like a puzzle maze. Thirty homes were pushed off their foundations. Three were reported gone down the river...One house was rammed by a large tree trunk, which had rushed down the river and lodged inside the house.*

"Clear Creek, that picturesque, tinkling addition to the decor of the Brookside Lodge as it flows through the dining room and chapel below, added its angry voice to the general uproar and swept away some 300 chairs and tables which have yet to be found".

FIGURE 5.23. Aerial view of the Front Street and Highway 1 intersection during the Christmas 1955 flood. Highway 17 was under construction at the time and what is now the Riverside Plaza in the middle of the photograph was completely inundated. The Cemetery on Ocean Street extension is on the left side of the photo. (Photo: Lee Blaisdell, courtesy of the Santa Cruz Beach Boardwalk Archives).

FIGURE 5.24. Virtually all of downtown Santa Cruz was under water during the Christmas floods of 1955 flood on the San Lorenzo. The Cathay Chop Suey restaurant was on the north end of Pacific Avenue. (Photo courtesy of the Santa Cruz Public Libraries).

FIGURE 5.25. 1955 flood waters moving down the northern end of Pacific Avenue. (Photo: Ed Webber, courtesy of the Santa Cruz Beach Boardwalk Archives).

In addition to the damage and disruption throughout the San Lorenzo Valley, as the river entered the city of Santa Cruz, it spread out across the now developed low-lying floodplain (*Figure 5.23*) and inundated much of the downtown area (*Figures 5.24, and 5.25*).

From the *Sentinel* of December 23, 1955):

"Santa Cruz braced itself for another night of

FIGURE 5.26. Panoramic view of flooding along the San Lorenzo River in the 1955 floods, looking at the intersection of Front Street and Laurel Street from approximately the location of the Warriors Arena. At this time there was no Laurel Street Bridge. (Photo courtesy of the University of California Santa Cruz Special Collections).

FIGURE 5.27. The lower San Lorenzo River with a former island during the 1955 flood. View is up Ocean Street from above Beach Flats. (Photo courtesy of the University of California Santa Cruz Special Collections).

wet terror after undergoing the greatest flood in the recorded history of the area last night. The still-rampaging San Lorenzo River roared from its banks at 9:30 last night in a destructive surge that drowned the city's central districts in up to 10 feet of water (Figures 5.26, 5.27 and 5.28).

FIGURE 5.28. Flooding along River Street in 1955, looking west towards the Garibaldi Villa Hotel, which was an active center of activity between 1894 and 1958. (Photo: Ed Webber, courtesy of the Santa Cruz Beach Boardwalk Archives).

"Pacific Avenue and the residential and industrial areas bordering the river were inundated by smashing torrents of water that tore up stores, shattered homes and forced hundreds upon hundreds of residents to flee to higher ground. A number of persons were reported swept from sight by raging currents that smashed toward the sea on both sides of the river (total death toll in the city was actually five persons).

"The city's water supply was cut off by the disastrous flood, and the only certain supply this afternoon was the 25 million gallons in the Bay Street reservoir.

"Santa Cruz auto row, which ranges from one end of flood-stricken Front Street to the other, was among the hardest hit businesses of the community-if not the hardest. One dealer, Palomar Garage, reported its entire stock of cars, approximately 50 used and 25 new, were either destroyed or badly damaged.

"Grim-faced Pacific Avenue merchants this morning supervised crews cleaning the swirling waters and slick mud from their stores... A check with a few of the merchants, bleary-eyed from the all night fight to hold out the flood waters, showed damage was severe throughout the downtown business district. Five feet of water eddied through the basement of the large store (Leask's Department Store), *and silt washed in by waters from the flooding San Lorenzo River covered the entire first floor.*

"For seven and a half hours, William Williams... shivered atop a telephone pole on Water Street until the flood subsided. When the flood came to the level of the car, Williams climbed on top of the hood of his car. Then with a sudden deluge of water the car toppled over and he was swept down Water Street. After being carried "quite a ways", he managed to get hold of a telephone pole. He was rescued at 5:10 a.m. today..."

Things were not going so well in Soquel either (*Santa Cruz Sentinel* from Christmas Day, 1955 by Wally Trabing):

"While Santa Cruz was stumbling under...its

flood troubles, the little community of Soquel was taking the beating of its life. Friday morning I waded in mud ankle-deep along Main Street on the west side of the Soquel Creek Bridge (See Figure 5.6) *and inspected the terrible damage that the uncontrolled waters dealt to the merchants. Thousands of dollars worth of merchandise was torn from shelves and racks, automobiles were swept helter skelter a block from the river and some have completely disappeared. Ed Peretto's variety store, Bill Finta's drug store, Willibanks Red and White Grocery Store were a complete tangled soggy loss within.*

"I saw abandoned homes along Porter and Walnut streets, homes once brightly decorated for Christmas, now desolate and water-soaked.... Soquel Creek, which draws residents to its sides because of its lazy pastoral gait, suddenly switched into a Mr. Hyde-like monster in the mountain darkness and lefts its banks Thursday night and Friday morning to tear savagely at the town's business district (Figure 5.29). *Soquel Fire Chief Roy Negro said last night that the flash flood caught many owners in their establishments. They took to the rooftops and attics and shivered helplessly through the night. He estimated that 36 to 40 homes were evacuated along the river in the Soquel area. The current was so powerful down Main and Porter streets that it battered automobiles about from one side of the street to the other...Soquel Lodge near the bridge lost five or six cabins. The Soquel Ford Garage, once filled with its heavy equipment and cars, was swept clean. 'It looks like an abandoned building now'.*

"Ed Hopkins, owner of a business on Main Street, said his place is destroyed. The walls are split. 'A 14-cubic foot refrigerator disappeared through the wall during the night and I can't find it', he said. Hopkins spent most of the period on his roof. George Vaughn parked his car near Izant's Hardware Store, a block from the river, and hasn't seen it since...The youth center behind the Soquel library was pushed off its foundation and jammed against the library.

Residents of the low land were to be allowed to return to their homes yesterday afternoon. Capitola itself suffered little damage. Its beaches look like a fallen for-

FIGURE 5.29. Damage from flooding in Soquel Village upstream from the Soquel Drive Bridge during the 1955 Christmas flood. (Photo courtesy of Carolyn Swift).

est. Amid the litter I saw beds, bits of clothing.. furniture, lawn chairs, and all types of household articles. Out in the breakers, the remains of a late model automobile, twisted and wrecked…"

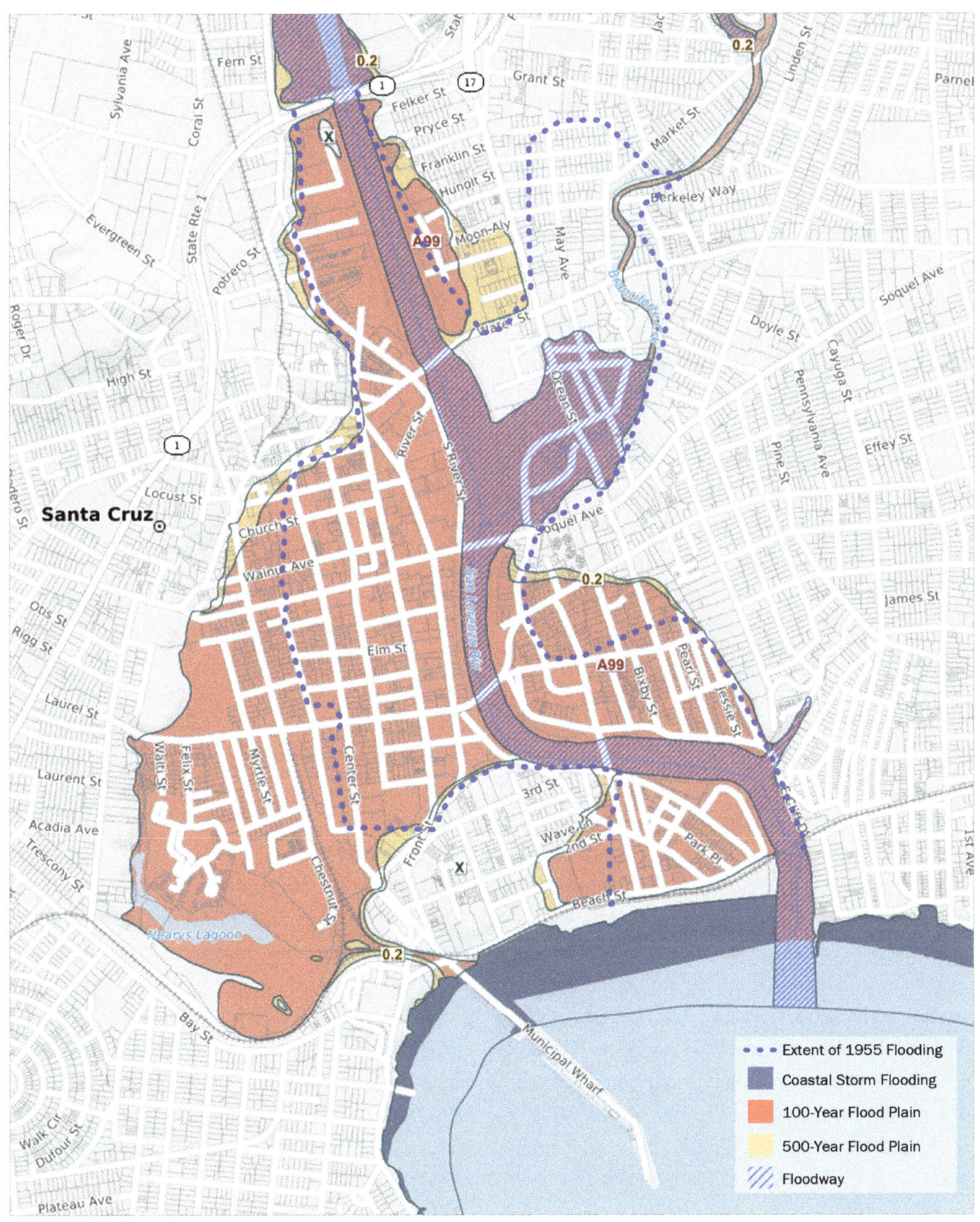

FIGURE 5.30. Map of the floodplain of the San Lorenzo River in downtown Santa Cruz and the area flooded during the 1955 flood. (Modified FEMA Flood Hazard Map).

Following the devastating Christmas flood of 1955, efforts were renewed and legislation was passed to undertake a flood control project on the lower San Lorenzo River as it flows through the city of Santa Cruz. As with many disasters, whether floods or earthquakes, there is always a flurry of legislative activity and no shortage of discussion, debate, and proposals to "*fix*" the problem or reduce future damage immediately after the event. Typically, this subsides after a few months and life returns to normal, with more having been said than done.

The bulkhead below Mission Hill (Bulkhead Street) built in the 1860s was the first attempt to protect the earliest downtown buildings constructed on the floodplain from being lost by bank erosion from further meandering of the river. Discussions and proposals continued through the remainder of the late 1800s, into the early 1900s and then reached a critical point with the flood of 1940. The Army Corps of Engineers and Congress began the process of planning, designing and appropriating funds for a flood control project through the city.

The Christmas flood of 1955 and the areas inundated (*Figure 5.30*) interrupted the planning process and led to a redesign to handle a larger flood. Engineering of any flood control project, whether a culvert, storm drain, levees or a dam, involves a certain amount of uncertainty. Normal practice is to use all of the years of stream flow records for the river in question (in this case, the San Lorenzo) and then complete a flood frequency analysis, which develops the probabilities of certain size floods occurring. This is a fairly standard process, but the reliability of what comes out the other end of the analysis, the 100-year flood flow for example, is very dependent upon how many years of stream flow records exist for the river, and how typical those years actually were. As was mentioned earlier, in the case of the San Lorenzo River, stream gauging or flow measurements had only begun in 1937, so that by the 1950s, there were only about 20 years of record. Calculating the 100-year flood, which the flood control project needed to accommodate, from only 20 years of river flow, involves a lot of extrapolation and therefore a considerable amount of uncertainty and guesswork. But, Santa Cruz has two more floods coming before the flood control project actually got underway *(See Figure 5.3)*.

February 19, 1958 Flood

Before the flood control channel and levees were completed, the heavy rains of February 1958 produced additional flooding within the city of Santa Cruz. Boulder Creek received 6.6 inches of rain in 24 hours causing the San Lorenzo to top its banks. February rainfall in Santa Cruz reached nearly 14 inches, one of the wettest Februarys on record. The Burnett Street area was the first to flood, but because the homes had been demolished to make way for a downtown redevelopment project, damage was minor. The flooding also extended onto Broadway, Laurel and Garfield streets. In Felton, residents of both Felton Grove and Gold Gulch were again evacuated as several cabins in each low-lying riverfront settlement were inundated.

April 2-3, 1958 Flood

Heavy rains continued into the spring in 1958 making this season one of the wettest in Santa Cruz County history. Rainfall totaled about 46 inches in Santa Cruz and nearly 87 inches in Boulder Creek by May. One hundred and twenty-five homes and 25 businesses suffered flood damage along the river downtown as car dealers along auto row (Front Street) were forced to move several hundred cars to higher ground.

"Rain-weary and flood-wary Santa Cruzans were well prepared as water spilled over the banks and into the traditional paths of previous high wa-

ters. Businessmen hoisted their stock from basements and first floors following a telephone warning... What probably saved the city from more extensive damage and heartbreak were the flood control and redevelopment projects which have removed many of the most susceptible homes and buildings from the water's path. Sections of Pacific Avenue, Front, Center, Barson, Spruce, Laurel and Cedar streets and Canfield Avenue were closed due to flooding" (Figures 5.31 and 5.32).

Potentially dangerous logjams at the Water Street, Soquel Avenue and Riverside Avenue bridges were avoided as cranes and other heavy equipment lifted, winched and pulled them apart to send the debris on its way downstream before serious logjams developed.

January 1967 Floods

The *Santa Cruz Sentinel* reported on January 17, 1967 that the "*residents of Felton Grove were slowly moving back into their homes today to begin the messy task of cleaning up from Tuesday's brief flood. Some 25 homes or more were either invaded or surrounded by the muddy waters of the San Lorenzo River...Many of the homes on Circle and River drives, whose owners have experienced floods as before, have been raised. These escaped the water which swept through the community up to four feet deep.*"

FIGURE 5.31. Flooding in Santa Cruz during the 1958 flood. View towards Front Street from near present-day River Street. (Photo courtesy of the University of California Santa Cruz Special Collections).

FIGURE 5.32. View to the Boardwalk from Ocean Street, which is underwater from the 1958 overflow of the San Lorenzo River. (Photo courtesy of the Santa Cruz Beach Boardwalk Archives).

Less than a week later, another storm hit the area:

"Mother Nature outdid herself in ending a six-week dry spell, releasing rain and winds which raised rivers to flood points, washed out roads, set loose landslides, knocked over trees... and caused extensive power and telephone outages....Main concern was in the San Lorenzo Valley... Felton Grove was awash and some people reportedly left their homes. There were also preparations for evacuation at Paradise Park and other low-lying sections. The creek through the closed Brookdale Lodge reportedly overflowed, causing some interior flooding.... Numerous slides were reported on such roads as Bonny Doon, Lompico, Bean Creek and Creek Drive."

January 16, 1973

Santa Cruz Sentinel headlines read *"Felton Homes Hit as Rain Swells River"*:

"Rains of cloudburst proportions lambasted the Santa Cruz Mountains during the night, filling the San Lorenzo River to flood stage and over and generally messing up county and state roads with slides and washouts. Some 25 homes in the Felton Grove area, just south of Felton, were evacuated about 10 a.m. as muddy, swirling water crept over the porches and into the living rooms.... The river oozed over its banks in Ben Lomond, flooding the recreation park area in town, covering the square dance and basketball areas".

The Big Storm of January 3-5, 1982

On the evening of January 3, a cold front from an Alaskan storm converged with a second storm originating in the warmer Hawaiian latitudes and sat directly over the central California coast for well over a day. Throughout the greater San Francisco Bay region, thousands of people vacated homes in hazardous areas, entire communities were isolated as roads were blocked, public water systems were destroyed or damaged, and power and telephone services were disrupted. Altogether, the storm damaged 6,300 homes, 1,500 businesses, and dozens of miles of roads, bridges and communication lines. Estimates for total storm damage compiled shortly after the event exceeded $280 million ($720 million in 2017 dollars). Lawsuits and claims filed against local governments reached $768 million (in 2017 dollars). Thirty-three people died as a result of the storm; 25 of these due to debris flows and landslides.

Intense rainfall lasted about 28 hours in the Santa Cruz Mountains with some locations recording one inch per hour for over eight hours. The downpour wasn't confined to the mountains; 8.23 inches of rain was dumped in Santa Cruz in 24 hours, exceeding any other 24-hour period of rainfall in the 90 years of record keeping. Ben Lomond got 11.50 inches in the same period, while Boulder Creek received 12.74 inches. The precipitation epicenter was in Lompico where 15.50 inches fell in 28 hours. Most of the San Lorenzo Valley had already been deluged with 33 to nearly 40 inches of rain in November and December of 1981 before the January 3-5, 1982 storm hit, so the soils were primed for failure. The additional January rainfall landed on saturated or nearly saturated ground so most of it turned to runoff, which rapidly reached the creeks and rivers. This excess water penetrating into the hillsides also weakened the soils and weathered bedrock producing hundreds of mudflows and landslides in the Santa Cruz Mountains as well as throughout the greater San Francisco Bay area.

On the morning of January 4, the County Communications Center began receiving the first of what were to become nearly 3,500 emergency calls by the end of the day. Initially it was landsliding and debris flows on a number of mountain roads that were the first obvious impacts of what was to become a 100-year rainstorm for much of the county. By noon on the 4th, the San Lorenzo River had topped its banks and entered the low-lying grassy area of San Lorenzo Park next to the Government Center (*Figure 5.33*). A deluge of calls reporting additional landslides, debris flows and flooding in the San Lorenzo Valley quickly followed.

FIGURE 5.33. In early January 1982 the San Lorenzo River overflowed onto the bench lands at San Lorenzo Park below the County Building. View downstream towards the footbridge from the Water Street Bridge.

By midnight, a logjam at the Soquel Drive Bridge in downtown Soquel extended over 300 feet upstream, and had diverted nearly the entire flow of the creek through the downtown business district (*Figure 5.34*). Creek waters came over the banks and into the Heart of Soquel Mobile Home Park so quickly that many residents didn't even realize there was a problem until water started to seep under their doors and their homes were already surrounded by knee-deep water. While explosives were considered to remove the pile of logs and debris, in a repeat of the 1955 logjam, fire officials opted instead for a crane from the county.

FIGURE 5.34. Damage and mud deposition in a mobile home park in downtown Soquel during the 1982 floods. (Photo courtesy of Bruce Richmond, U.S. Geological Survey).

Unfortunately, the crane broke down en-route and never made it. The raging waters took a new course through the Old Mill and Heart of Soquel Mobile Home Parks, past the Memorial Library and directly through the main intersection of Porter and Soquel Drive where it reached a depth of five feet. Twenty-two of 39 mobile homes as well as automobiles left behind were listed as damaged beyond repair as up to five feet of mud was deposited in the parks. The trailers were floated off their piers, tipped on their sides, filled with debris, or buried *(See Figure 5.4*). Floodwaters and the debris they left behind damaged the firehouse, post office, library and Grange Hall, as well as 58 businesses and 21 homes *(See Figure 5.8)*.

Several relatively new homes had been permitted on the floodplain along Wharf Road next to Soquel Creek with the condition that they were elevated to avoid future flood damage. Only carports were allowed at ground (flood) level. Following construction, however, external walls were built and those open carports became bedrooms and other habitable areas. As expected, the creek overtopped its banks and flooded these first

FIGURE 5.35. Homes along Wharf Road in the village of Soquel were built on the low floodplain of Soquel Creek and were flooded up to the mud line on the side of this home.

floor structures leaving several feet of mud and considerable damage behind (*Figure 5.35*).

The size of the flood and the formation of the logjam at the Soquel Drive Bridge were almost identical to the conditions in 1955. A resident who had also experienced the 1955 flood stated that the 1982 flood, though slightly lower in elevation, was much worse because "*it hit harder and quicker*", and because so much growth had occurred in the business district since the 1955 event. Although a number of flood control projects have been discussed and considered for protecting Soquel, the village is still completely unprotected. Because of the low-lying nature of the community, protection would be very difficult to achieve. The 1982 flood, which was so damaging to downtown Soquel *(See Figure 5.8)*, was not the 100-year event but statistically only a 15-year flood. Floods of this size and larger have and will occur relatively frequently, in fact about every 15 years on average.

While development along Aptos Creek is much less extensive than that of Soquel, the homes between the freeway and the Spreckels Drive Bridge, most of which had either been rebuilt after 1955 or constructed on stream side fill brought in after the 1955 flood, suffered heavy damage. Aptos Creek, as with most streams that have built floodplains, form windy or sinuous channels. Following the 1955 floods, the creek channel was constricted and straightened by fill, followed by the construction of a number of creekside homes along Moosehead and Spreckels Drives. In January 1982, the creek cut a meandering or curving path to the sea, after it passed under State Highway 1, which quickly eroded the fill and undermined the foundations of the creek bank homes (*Figure 5.36*). At least seven homes sustained major dam-

FIGURE 5.36. A number of homes were built along the bank of Aptos Creek on Moosehead Drive and suffered major damage and collapse during high water in January 1982.

FIGURE 5.37. At least some of these homes were built on fill next to Aptos Creek placed after the last major flood, which was eroded in the 1982 floods.

age (*Figure 5.37*). Two homes broke in half, with the detached part of one floating downstream. One house completely collapsed into the creek with a portion of it ending up on the Spreckels Drive Bridge (*Figure 5.38*), while the rest floated to the bay. In contrast to the village of Soquel, where homes and businesses are built on the floodplain, along Aptos Creek, these homes were built on the stream bank, directly in

FIGURE 5.38. This home on Moosehead Drive was floated downstream and got caught on the bridge at Spreckels Drive during the 1982 floods on Aptos Creek.

the floodway, and in many cases, on unconsolidated fill placed in the stream channel. In short, these are not safe places to build. Yet, following the 1982 flood, the creek channel was again straightened and the homes were rebuilt again on fill in the same flood-prone location.

The flood control levees along the lower San Lorenzo barely saved the city of Santa Cruz from inundation. While the river flowed onto the benchlands of San Lorenzo Park *(See Figure 5.33)*, this low area was planned for overflow, and other than some sand and debris deposited, there was no damage. Elsewhere the river rose to within three feet of the top of the levees, which is about as close as engineers would like to ever see. Severe scour by the high flows undermined part of the Soquel Avenue Bridge which then collapsed (*Figure 5.39*), in the process breaking all of the phone lines connecting east and west sides of town (this was in the days before cell phones). The Riverside Avenue Bridge also was damaged by settlement due to scour. Cranes were busy throughout the night of January 4-5, keeping logjams from forming at the low arched Riverside Avenue Bridge *(See Figure 5.5)* and also the upstream Highway 1 Bridge.

Upstream in the San Lorenzo Valley, the residents were not as fortunate. There are no flood control projects and overflow was widespread, typically in the same areas that had been flooded many times in the past. Between Felton and Santa Cruz, individual developments adjacent to the river, including Felton Grove, Gold Gulch and Paradise Park were all severely damaged. Felton Grove is situated on a very low area of the floodplain and river overflow had inundated the area closest to the river three times in the previous twelve years (1969, 1973, 1978 – *Figure 5.40*) and 14 times in the previous 46 years. While we tend to hear about the "*100-year floodplain*" as the area adjacent to a river that would flood once every 100 years on average, the cabins and homes in Felton Grove closest to the river are on the "*3-year floodplain,*" or are flooded about every three years on average. These aren't very good odds. Virtually all of the houses in Felton Grove were flooded by high water on January 4, 1982 as flood waters extended all the way to Graham Hill Road (*Figure 5.41*). Water levels reached three to six feet in many homes and left three to four feet of sediment behind, in homes and in automobiles (*Figures 5.42, 5.43 and 5.44*). The residents dug themselves out again and have reoccupied their homes, although some were eventually elevated.

FIGURE 5.39. The concrete piers supporting the Soquel Avenue Bridge across the San Lorenzo River in downtown Santa Cruz were undermined by scour and settled in the 1982 floods, leading to partial collapse of the bridge. Note debris left by high water on the collapsed section of roadway.

FIGURE 5.40. Some homes in Felton Grove were flooded in 1978 with high water on this house recorded by the debris line next to the stone chimney.

FIGURE 5.41. The 1982 overflow of the San Lorenzo River at the Covered Bridge in Felton Grove reached up Covered Bridge Road all the way to Graham Hill Road.

In Gold Gulch, about a mile south of Felton, a group of about 25 homes was inundated and damaged. In 1955, 30 homes in this area were reported pushed off their foundations. Maximum water depths near the river reached 12 feet in one two-story home.

Overflow from Bull Creek inundated some businesses in Felton, including the firehouse, which had a

FIGURE 5.42. This Felton Grove house adjacent to the covered bridge (in right background) was surrounded by sediment following the 1982 floods.

foot of water running through it. Between Felton and Ben Lomond, 60 to 70 homes and other structures along the river were flooded as was the Mill Street business section of Ben Lomond. The effects of flooding in the valley were felt in the city of Santa Cruz as overflow from Loch Lomond Reservoir broke the city waterline requiring 70,000 users to ration water until repairs could

FIGURE 5.43. This Felton Grove home and the VW bus were buried in sediment by the 1982 floods. Note the high water mark on the white frame near the top of the house windows.

FIGURE 5.44. San Lorenzo River floodwaters flowed over the top of these cars in Felton Grove in January 1982 and left them full of sand.

be made. It's always difficult to understand why there are water shortages during a flood but damage to water lines and lost electrical power frequently accompany large floods.

An equally damaging impact of the 1982 rainstorms was the slope failure, primarily landslides and debris flows. The Love Creek disaster is most significant but this was only the largest of many throughout the Santa Cruz Mountains and the greater San Francisco Bay area. These are discussed in Chapter 6 on Landslides, Mudflows and Rockfalls.

1986 Storms and Floods

The winter of 1986 was one of well above average precipitation and added to the incidents of flooding and slope failures experienced in the wet winters of 1982 and 1984. The most intense period of rainfall occurred in mid-February and was labeled the Valentine's Day storm. Ben Lomond and Boulder Creek each recorded just over 26 inches of rain in a week. While most of the serious problems, including countless road closures, were due to landsliding and debris flows, one area in particular was hit by both slope failures and flooding. Eureka Canyon Road, which follows Corralitos Creek as it heads into the Santa Cruz Mountains east of Corralitos, was hit extremely hard as the intense rain produced a damaging combination of landslides, falling trees and stream erosion which washed out much of Eureka Canyon Road and destroyed cabins and motor vehicles (*Figures 5.45 and 5.46*)

MONTEREY COUNTY FLOODING

The rivers and their watersheds in Monterey County are quite different that those of Santa Cruz County. The landscape undergoes a dramatic change at about the Pajaro River, which is the boundary between the two counties, and sep-

FIGURE 5.45. Flooding along Eureka Canyon in 1986 ripped out the only access road.

FIGURE 5.46. Floodwaters moved so rapidly down the steep Eureka Canyon stream that homeowners didn't have time to move vehicles. Large boulders were transported downstream adding to the damage.

arates the steep, mostly redwood-covered Santa Cruz Mountains to the northwest, from the more arid, mostly chaparral-covered Santa Lucias to the southeast. The drainage basins of the San Lorenzo River and Soquel Creek are quite steep, short and small (138 square miles and 42 square miles, respectively) such that flooding occurs rapidly after heavy rainfall but then doesn't last long. The Pajaro River (1,300 square mile drainage basin), Salinas River (4,160 square miles), and Carmel River (250 square miles), are considerably larger but receive much less rainfall than the river basins to the north. As a result, large floods require much more sustained rainfall and take much longer to move downstream. Additionally, all three of these Monterey County rivers draining into the Monterey Bay area have very broad alluvial valleys or floodplains such that flood waters can spread out and cover very large areas – most of it agricultural land – when the banks are overtopped.

The most complete record of historical flooding in Monterey County has been compiled by the Monterey County Water Resources Agency and is available on line at: http://www.co.monterey.ca.us/government/government-links/water-resources-agency/programs/floodplain-management/historical-flooding#wra

The following historic flooding accounts are drawn primarily from that source.

March 1911 Flood

Heavy rainfall in early March caused flooding throughout central California. The headline in the March 8, 1911, issue of the Salinas Daily Index described the details of storm and flood conditions along the Salinas River at Spreckels:

"Disastrous effects of the storm in the Salinas Valley is unprecedented.

"This storm was the most disastrous in the history of Monterey County and the damaged property is unprecedented. It is reported that more than 2,000 acres of valuable farming land has been destroyed along the course of the Salinas River by the cutting away of the banks of that stream, which is now a raging torrent, freighted with debris, from its source to its mouth on the Bay of Monterey, near Moss Landing... At 10 o'clock the river was said to be higher than at any time since the winter of 1862".

The Spreckels area (a few miles south of Salinas and adjacent to the Salinas River) was heavily damaged by river overflow:

"At Spreckels, all the lowlands are flooded and the water comes to within thirty feet of the end of the (sugar) *factory, which is protected by a heavy rock embankment. The river is nearly a mile wide at some points there.*

"The electric light plant and the pumping plant, as well as two large oil tanks near the factory, are half submerged. The No. 2 tank has been torn loose... Barns and outbuildings and farmhouses all along the river bottom south of Spreckels are under water, and tops of a few being all that remain. Everything not securely anchored has been swept away."

Traffic on the coast line of the Southern Pacific Railroad was halted as 300 feet of trestle was washed out about 35 miles south of Salinas. Another washout between Castroville and Monterey shut down traffic on the Monterey branch. The Carmel River also went on a rampage, sweeping away four cabins built near the river and used by Salinas residents for summer homes.

January 1914 Floods

The storms of January 1914 did significant damage throughout Monterey County (Salinas Daily Index):

"Flood conditions prevailed today everywhere throughout the Salinas Valley. Bridges have been carried away, the railroad trains tied up, telephone and telegraph service interrupted, and inestimable damage done as a result of the torrential rains of Saturday night and Sunday. Salinas has been isolated as far as communications south to Soledad and north to Cas-

FIGURE 5.47. Postcard of U.S. Mail delivery by cable car after the 1914 flood on the Salinas River destroyed 500 feet of the bridge into King City. (Photo courtesy of Carol Harrington).

troville is concerned..

"Monterey County has suffered an enormous loss through the damage and destruction of bridges. Passengers arriving from Soledad and Gonzales say private reports received at those places indicate the loss of all the bridges south of Chualar. The Bradley, San Ardo, San Lucas, King City, and Soledad bridges are gone (Figure 5.47). Two spans of the Gonzales Bridge have gone out. At Chualar, one end of the bridge has sunk two feet and is one foot out of line. At Gonzales, the people were this morning constructing a cable line over which to send food and supplies on the other side".

December 1931 Flood

A Christmas storm in 1931 brought flood conditions to many portions of Monterey County. Precipitation was dramatic; on the Carmel River the San Clemente Dam overflowed capacity. As noted in a December issue of the *Monterey Peninsula Herald*: "*Fed by storm swollen streams, San Clemente Dam staged the most sensational rise in history last night, climbing 25 feet in 15 hours.*" The storm lasted for 5 days and brought damage to Carmel Valley, Big Sur, and the Monterey area.

February 1938 Flood

In February 1938, the Salinas River again flooded. The headline in the *Salinas Index-Journal* of February 12 stated: "*No, not the Mississippi-just the Salinas River.*" Conditions in the county were serious. At a dozen points along the 70-mile river front from King City to the coast the churning water brought to an unprecedented high by the heavy rains in the mountains and valley brought damage and halted traffic on bridges and roads and marooned an estimated 60 families along the River Road on the west side of the river.

Going out with a roar that was hardly heard above the driving rain and lashing flood waters of the Salinas River, 208 feet (two spans) of the Soledad bridge on U.S. Highway 101 were swept downstream at 9:15 p.m. Friday night, adding wreckage to the swollen river, which by Saturday afternoon appeared to have reached as high as the most severe floods in the valley in years.

Floods of 1940-1941

The heavy rainfall of the winter of 1940-1941 produced flood conditions in several areas of Monterey County. At the Trescony Ranch in the San Lucas district, 23 inches of rain had fallen that year to make it the wettest sustained period in history and the largest amount of rainfall for any season since 1890. Streets were flooded at Soledad, and old-timers said that the water was the highest since 1910.

The River Road, a half-mile south of Spreckels and southward, was flooded and motorists were advised not to attempt to negotiate it. The Arroyo Seco road was closed to traffic, as was the Pinnacles route out of Soledad. A washout also blocked the Jamesburg road in the upper Carmel Valley. Both the piers and the foundations of the approaches to the Toro Creek Bridge were washed out by floodwaters, making the span unsafe for traffic.

January 1943 Flood

A dramatic storm hit the Monterey Peninsula in January 1943, bringing flood conditions to coastal areas of Monterey County. The *Monterey Peninsula Herald* of January 22 described the event:

"A downpour of cloudburst proportion flooded upper reaches of the Carmel Valley during Monterey Peninsula's worst storm in a quarter century it was revealed as reports began coming in from the outlying regions today. While counting the storm damage continued to occupy local residents, it was reported that 5.50 inches of rain had fallen at San Clemente Dam in the 48-hour period ending at 9 a.m. today.

"During most of yesterday, over 6 feet of water was thundering over the spillway at the rate of 8,000 cu. ft. per second, enough to fill the dam 7 times each day. It is estimated by water company engineers that enough water passed over the spillway during the storm to supply Monterey Peninsula for the next four years".

Seventy-two years later, in 2015, the San Clemente Dam was demolished and became the largest dam removal project in California history. The dam was deemed no longer seismically safe and the reservoir had accumulated 2.5 million cubic yards of sediment such that it had virtually no remaining water storage capacity left.

January 1952 Flood

As noted in the Salinas Californian, 1952 was another significant flood year in Monterey County:

"The rampaging Salinas River, swelled by 6 days of heavy rain, today had left its banks, flooded Spreckels Junction and forced evacuation by boat of several families in that area and also in Salinas on East John Street. The Salinas-Monterey Highway was closed at Spreckels Junction Bridge and probably will not be opened until tomorrow…

"Old-timers said the river was the highest it has been since the 1911 flood, and reports this morning from King City said that the stream in that area was rough and high. A crest of the river was expected today when water from yesterday's rain in the mountains reaches this area…"

April 1958 Flood

The torrential rains of early April 1958 brought flood conditions to numerous counties in northern California. Monterey County was no exception, as reported in the Salinas Californian:

"Flood water swept through Monterey County today as streams in the Salinas and Carmel Valley watersheds overflowed their banks, closed roads, endangered residents, drowning poultry, and damaging homes. The Carmel River has gone over its bank, flooding numerous home tracks bordering the river the length of the valley.

"The Nacimiento Dam was reported filled and water is being released slowly to take off the peak. Nearly 3-1/2 inches of rain in 24 hours in the Arroyo Seco has turned the placid stream into a raging torrent ripping through summer cabin sites on its way to the already swollen Salinas River. In the Greenfield area, a marooned family was rescued by Army helicopters. The Salinas River has overflowed its banks in numerous places, causing the closing of the River and East Garrison Roads. Water may overflow the Salinas-Monterey Highway as a result of the record flow in the Arroyo Seco River.

"San Lorenzo Creek overflowed its banks in King City and spread through a chicken ranch, drowning 23,000 birds".

Major Floods of 1969

The year 1969 was perhaps the most severe flood year in the recent history of Monterey County. There were two distinct floods, one at the end of January and one at the end of February, and both of these led to Monterey County being declared a disaster area. Both the Salinas and Carmel rivers went on a rampage during each of these floods. Damage from the storms was extremely costly.

The Salinas Californian described the flood conditions within the county:

"The Salinas River cut a multi-million dollar swath of damage through the Salinas Valley from Bradley to the Pacific Ocean today. The Valley has been awash in what County Water Engineer Loran Bunte calls the 100-year flood since Saturday evening. A flood crest only slightly lower than that which passed Spreckels at 40,000 cu. ft./sec. early this morning, is rolling up river from King City this afternoon.

"Monterey County Administrator and Civil Defense Director Walter Mansfield declared the county a disaster area Sunday. His declaration triggers the mechanism through which the county may be compensated with federal funds for public facilities damaged by the flood. Salinas Valley agriculture, which sustained a $3,755,000 loss in the 1966 flood, will almost certainly be hit harder this year".

One month later, the Salinas River again flooded with more damage left behind, as reported in the Salinas Californian on February 26, 1969:

"The Salinas River, fast, deep and a mile wide, flowed as a flood crest through the Salinas Valley this morning, cutting a swath of muddy destruction. Route 1 was closed at 10:30 a.m. at Twin Bridges near Nashua Road as the river's crest surged toward the ocean, overflowing the highway and drowning the artichoke field delta around Mulligan Hill.

"The city of Salinas, which underwent some anxious moments fretting about the possibility of urban flooding last night, remained high and dry as the crest passed".

In September of 1968 I just had moved to Santa Cruz to accept an Assistant Professor position at the new University of California campus. A few months later on a dark and very stormy night in February 1969, while driving home from dinner in Monterey, I vividly recall that the fields on both sides of Highway 1 for a considerable distance on either side of the Salinas River crossing were completely flooded. The narrow band of asphalt was the only surface that wasn't covered with floodwater and with water lapping up on the edges of the road it felt like driving over a floating bridge.

February 1978 Floods

In 1978, flood conditions again occurred in many areas of Monterey County, as noted in the Salinas Californian on February 13:

"Pounding weekend rains have left Salinas Valley farmers looking at an estimated $20,000,000 in flood damages today. Damage was concentrated along the banks of the Salinas River from San Ardo out to the sea.

"More than 20,000 of the valley's 200,000 irrigated acres of land were covered with overflow waters from the Salinas River at some point Saturday or yesterday. As much as 1,000 acres of the valleys prime farmlands could be flooded beyond agricultural use this year.

"The assessment of damages, exceeding those of even the valley's 1969 flood... Damage would have been far more severe if not for the flood control capacities of both Nacimiento and San Antonio dams... Two dams, almost bone dry two months ago, were holding 290,000 acre-feet of water at Nacimiento and 137,000 acre feet at San Antonio this afternoon. That puts Nacimiento at peak holding capacity already.."

January and March 1995 Floods

Monterey County experienced prolonged precipitation in January 1995 resulting in extensive flooding throughout the region. Most river valleys were affected, with major damage experienced in the Pajaro Valley and Carmel Valley. On January 9 and 10, 1995, Monterey County received up to six inches of rainfall, ranking it as a 10 – to 20-year event. Five localized areas within the Carmel Valley area were significantly affected by downstream river flooding: Camp Stephani, the Robles Del Rio area of Carmel Valley village, the area adjacent to the Schulte Road Bridge, the Rio Road area adjacent to Highway 1, and Mission Fields, with damages to 125 residences as well as public facilities and utilities.

From March 10 – 13, 1995, Monterey County experienced a second major winter storm which resulted in sustained precipitation falling on already-saturated watersheds. Devastating flooding occurred throughout the county, particularly along the Carmel, Arroyo Seco, Salinas, and Pajaro rivers. Damage was extensive throughout the county with virtually every community affected. Pajaro, Castroville, Mission Fields, Carmel Valley, Cachagua, Carmel Highlands, Spreckels, and Big Sur sustained devastating damage. Over 1,500 residences were damaged, including 60 homes, which were declared uninhabitable. In addition, an estimated 110 businesses were damaged, and the tourism industry sustained substantial losses for a period of several months.

In all, over 11,000 people were directly affected, and major portions of the county's agricultural lands subjected to widespread damage. In particular, flooding of the Salinas, Carmel, and Pajaro Rivers forced mass evacuations in San Ardo, King City, Greenfield, Soledad, Gonzales, Chualar, Spreckels, the River Road area, parts of Salinas, Castroville, Moss Landing, Pajaro, and the Carmel Valley.

On March 13, at the height of the flood, 63 roads and 15 bridges were closed, including the Highway 1 Bridge over the Carmel River, which resulted in the complete elimination of access to portions of Carmel Valley, Carmel Highlands, and Big Sur for a period of several days. Public and private water systems were also damaged, affecting approximately 3,500 homes and businesses; many residents were without domestic water service for extended periods. Sewage treatment facilities and private septic systems along all three major rivers (Carmel, Salinas, and Pajaro) were flooded and untreated sewage was released into the rivers.

February 1998 Storms

In February 1998, a series of El Niño winter storms hit various parts of California, including Monterey County. Close timing of the rainfall events contributed to intense flooding, with heavy rain continually hitting ground that was still saturated from the previous rains. An estimated 50 roads and highways were closed or restricted, in most cases due to washouts, landslides, and mudslides. Several communities were evacuated, particularly the entire town of Pajaro, across the Pajaro River from Watsonville, all residents of the Sherwood Lake Mobile Home Park near Carr Lake in Salinas, and portions of Bolsa Knolls and Toro Estates. Drinking water quality warnings remained in effect for certain areas for some time afterward. By the end of the first week of February, at least 6,600 homes and businesses had been without power for varying periods of time. Countywide, losses resulting from the February 1998 events were estimated at over $58 million (2017 dollars), with agriculture-related losses totaling nearly $11 million (2017 dollars) and involving approximately 29,000 damaged acres.

SOME FINAL THOUGHTS ON FLOODING

While the residents of the central coast have seen more droughts than floods in recent years, the above history of intense rainfall and flooding is a reminder that we are never far away from another wet winter, at least historically.

Many of the cities and towns in the region were built all or in part on river floodplains and these flat agricultural areas have been formed by thousands of years of overflow and sediment deposition. These same floods created the fertile soils that support an immense agricultural industry providing vegetables to the entire nation. While levees have been built along some of the streams draining into the bay, and have proven effective for years in some cases, it is important to remember that all flood control stops somewhere. We cannot afford to build levees or dams to control every extreme flood. As the climate changes, we may well see more concentrated winter rainfall and more frequent and larger floods. Recognizing the risks of where we have built is the first step; how we respond to those risks as individuals and communities is more challenging, and historically our approach has been to build higher levees rather than relocate communities. All flood protection ends somewhere and we need to learn from the floods of the past.

Chapter 6

Landslides, Mudflows and Rockfalls

Well the landslide will bring you down...
– Stevie Nicks

INTRODUCTION

With increasing frequency, the sounds of snapping two-by-fours, shattering glass, and the cries of the suffering homeowner, and ultimately the arguments of the lawyers in court, have been heard as an accompaniment to an old and common geologic process-the landslide. Not long ago, an entire hillside could slide into a canyon or a slab of sea cliff would collapse into the waves and no one would much care. Now, however, as the hillsides of the Santa Cruz Mountains have become increasingly popular to homebuilders and homeowners, and many of the bluffs along the northern shoreline of the bay have been almost completely developed, slope failures of this sort have become frightening, devastating and expensive events. Along the steep Big Sur coast, massive landslides and extended road closures have become a way of life in winters with heavy rainfall for Caltrans and the scattered homes and businesses along this rugged coast *(Figure 6.1)*.

FIGURE 6.1. A very large landslide induced by high rainfall during the winter of 1983 closed Highway 1 in Big Sur for months. Note cars along the roadway for scale.

Landslides are just one of a number of a types of downslope movements or slope failures consisting of soil, rock and other materials such as trees or vegetation. These hillside failures are driven downhill by gravity and are often initiated or accelerated by excess rainfall or water, or less frequently, earthquakes. Some slides may involve only a truckload of wet soil or rock, while in other cases, deep failure may involve entire hillsides, hundreds of thousands of cubic yards of earth and rock, and dozens of homes. Some types of slope failure take place very slowly, moving only inches per day, while others, such as mudflows, move faster than any of us can run. The threat and risks vary then as to the amount and type of material moving towards your home or car or you. None of these events are going to be beneficial but some events may be far more destructive and dangerous than others.

FIGURE 6.2. Hundreds of debris or mudflows occurred throughout the Santa Cruz Mountains in early January 1982 as a result of the 100-year 24-hour rainstorm. This debris flow blocked Smith Grade in Bonny Doon for several days.

FIGURE 6.3. A very small debris flow on a hillside in Scotts Valley in January 1982 pushed this house completely off its foundation.

While not always obvious in the Santa Cruz Mountains because of the forest cover, landslides and other types of downslope movements are a widespread phenomena affecting the shape of the landscape we see in central California. For many, it is only when we see rocks or mud falling or oozing onto the roadway and blocking traffic that we even become aware of this all too common process *(Figure 6.2)*. Because of the high rainfall in our local mountains, 50 or 60 inches in most years, the scars left behind by slope failures are quickly vegetated and within a few years are usually unrecognizable. Even to the trained geologist, evidence for previous landslides or mudflows is not always obvious or apparent. What is more challenging in assessing the stability of a hillside is whether the area most likely to fail during the next intense storm is an area that has previously failed, or the adjacent area with the same materials and slope, which hasn't yet failed.

The more years that have passed since the hillside has slid or slipped, the more the landforms and vegetation conspire to heal the land and cover the evidence. This may be beneficial to those trying to sell mountainside property, but in the long run information of this sort on areas prone to failure is important to consider before buying or building in steep terrain. Having a modest sized 50 cubic yard debris flow (roughly 5 dump truck loads) collide with a typical wood frame dwelling is like driving that same truck into it at 15-20 miles per hour *(Figure 6.3)*. Most homes don't survive impacts of this sort very well and it's advisable to do everything we can to avoid this happening. Unfortunately, as with most geologic hazards or disasters, events like this occur irregularly and somewhat infrequently, and the memory of these usually fades fairly quickly. We often tend to share a collective amnesia for natural disasters and try to get over them and move on as quickly as possible.

For those just moving to the Monterey Bay area, there is no memory and little appreciation for the active geologic nature of the landscape in this part of the world. There is the bay, the beaches and those beautiful redwoods and rivers, which are the magnets attracting newcomers here. Why worry about things you can't even imagine or envision?

Whether you live here now or are considering a move, there are several different types of hillside failures that affect the steeper slopes of the Santa Cruz Mountains and that are of concern to us as residents, wannabe residents or commuters through any of the mountain passes to the employment in the valley on the other side.

FIGURE 6.4. Rocks falls of large joint-bounded blocks of Purisima Formation are extremely common along Depot Hill in Capitola.

ROCKFALLS

Rockfalls are perhaps the simplest type of slope failure to identify and observe and also the easiest to understand. These events occur when loose or weak rock in a sea cliff or along a steep road cut or canyon wall breaks loose and falls, rolls, or slides downhill. This process is very common along the high and near vertical sea cliffs of northern Monterey Bay from Natural Bridges to Aptos, particularly between 41st Avenue and New Brighton State Beach *(Figure 6.4)*. Much of the rock along these cliffs is relatively weak sandstone and mudstone and, due either to undercutting or impact by waves, excess rainfall, or shaking during an earthquake, the rock breaks loose along weakness planes and falls to the beach below. This can be a serious hazard if you happen to be either walking or sitting on the beach beneath these cliffs. Even a casual visitor would probably recognize the large rocks at the base of the cliff and hopefully have the common sense not to spread out their towel among the rocks.

A much greater concern exists for those who own homes on the cliff top who are gradually losing their back yards and patios with each chunk of rock that breaks loose. Depot Hill, between Capitola and New Brighton State Beach, is a good example of this problem *(Figure 6.5)*, and one that has no easy, inexpensive or long-term solution, although many have been proposed over the years. As long as waves strike the base of the cliff, and it continues to rain, failure of the rock in the cliff is going to continue to occur and threaten the homes and utilities along the cliff top.

On October 17,1989, we learned that large earthquakes can also trigger sea cliff failures. Seismic shaking during the Loma Prieta earthquake produced bluff failures from San Francisco to central Monterey Bay. One of the seven deaths in Santa Cruz County during that earthquake occurred at Bonny Doon Beach along the north coast when a sunbather was buried under a rock fall from the overhanging cliff. The bluffs throughout Capitola and Rio del Mar also experienced widespread collapse due to the severe shaking *(Figure 6.6)*. Cracks appeared at the bluff top beneath foundations and patios, large masses of rock and soil broke loose and slid or fell to the beach below, some of this blocking access to and damaging homes and burying cars at the base of the bluff in places like Pot Belly Beach and Las Olas Drive just south of New Brighton State Beach *(Figure 6.7)*. Six apartment units on Depot Hill in Capitola had to be demolished as the concrete caisson support system was undermined by cliff failure, which led to cracking of the foundations and walls *(See Figure 4.22)*. Three other bluff top homes were also ultimately demolished as their ocean view sites were cracked beyond repair by the shaking accompanying the earthquake. It's important to recall that the epicenter of that shock was only a few miles inland from these sites and that shaking was intense.

FIGURE 6.5. These apartments on Depot Hill in Capitola were built very close to the cliff edge in the 1960s and continued cliff failure progressively undermined the foundation of the apartments and the adjacent house. The house was ultimately moved to a separate property while the apartments were subsequently supported with a system of concrete caissons, which provided temporary support until the Loma Prieta earthquake.

Rockfalls are seen during virtually every winter along many of the roads that traverse the Santa Cruz Mountains, specifically, Highway 9, Highway 17, Old San Jose Road, Chittenden Pass and Hecker Pass, to name the major commuter arteries. Sometimes a lane is closed, and traffic is slowed down, and within a few days things return to normal. The Loma Prieta earthquake created some very large rock falls, in particular at the Laurel Curve, which closed Highway 17 for days *(Figure 6.8)*. While no

FIGURE 6.6. Bluff failure from seismic shaking during the 1989 Loma Prieta earthquake required the construction of large retaining walls below these homes in Aptos Seascape.

FIGURE 6.7. Cliff failure during the Loma Prieta earthquake partially buried this car and cut off access to Las Olas Drive immediately north of Seacliff State Beach.

FIGURE 6.8. The Laurel Curve rockfall along Highway 17 was one of many slope failures from the Loma Prieta earthquake and closed two lanes of the highway for weeks.

one was injured in this massive failure, there is a touching story of a father who commutes daily over the hill for an evening job. That particular afternoon, his young daughter asked him for one more hug before he drove off to work. He saw the slide take place just as he approached the curve and skidded to a halt, just missing being buried by the rock and debris.

Highway 1 at Waddell Bluffs, just south of Año Nuevo and the San Mateo County line, is another site where the weak rock in the steep 300-foot high bluffs continues to fail and then roll and slide downhill towards the highway. The California Department of Transportation (CalTrans) built a ditch alongside the road to catch the rolling and falling rocks; sometimes this works, but historically if the ditch hasn't been emptied for a while, large rocks frequently would bound out onto Highway 1. Despite the warning signs at either end of this hazardous stretch of highway asking you to "*Watch for Rocks,*" there is no follow-up message advising you what to do if you see a rock racing downhill towards your car. Vehicles have been damaged and waylaid by colliding with rocks that have either come to rest on the roadway or are still moving across the road *(Figure 6.9)*. A man lost his life at Waddell Bluffs in 1982 when a 200-pound rock rolled downslope, through the trench, and became airborne before breaking through the front window of a truck. Following a lawsuit against Caltrans, the maintenance of the trench has improved with more timely removal of the rocks and debris and also construction of a heavy cable barrier to prevent rocks from entering the highway.

FIGURE 6.9. State Highway One was built along the base of Waddell Bluffs in 1946 and has had to deal with regular rockfalls ever since, which often bounced onto the highway as can be seen in this photograph. After a fatal accident and lawsuit the California Department of Transportation installed a barrier next to the roadway to catch falling rocks before they reached the roadway.

LANDSLIDES AND SLUMPS

Landslides and slumps may include rock, soil, or a mixture of the two, as well as anything on the overlying landscape, including trees, houses, roads or cars. They can be small or very large, and are usually distinguished by relatively slow movement. These events involve a mass of material, which fails along a plane or surface and can move as a relatively intact mass but can also break apart as it moves downhill. If the failure surface is smooth or planar and the material is relatively undisturbed, it is called a slide because the material glides slowly or slides downhill. More commonly, the failed hillside material breaks up into a series of separate blocks and is distorted and rotates as it moves downhill, and is called a slump. Slides and slumps can be a few feet to hundreds of feet thick and may involve dozens of acres of hillside, leaving behind lots of deep cracks, and a jumbled mass of soil and rock.

The Santa Cruz Mountains are prone to landsliding and slumping *(See Figure 2.6)* for several important reasons: the rocks and soils are young geologically speaking and relatively weak; the slopes are often steep, giving ample opportunity for gravity to go into action; there is a long history of earthquakes which have shaken and weakened the rocks repeatedly; and the rainfall is very high during many winters, at least historically, such that plenty of water is available to further destabilize the slope by weakening the rocks and soils further. Most hillsides fail in the winter months when there has been a good amount of rainfall.

Some of the larger slide areas in the Santa Cruz Mountains have their own identities and names because they continue to plague roads or developments during particularly wet winters, the Blue Slide along Highway 9 between Felton and Ben Lomond, for example. Another example with a long history was the Mt. Hermon slide. This site of repeated failure involved a large hillside above Bean Creek along the old Mt. Hermon Road between Scotts Valley and Felton. Activity in the area had been documented back into the 1950s either following moderate earthquakes on during heavy winter rains. The large sand quarry along this hillside with its associated wash water basins, heavy equipment operation and vibration, as well as earthmoving activities combined with the unfavorable geology, which consisted of layers of rock tilted downhill towards Bean Creek, combined to produce intermittent movement of a few feet at a time. The result was uplift and displacement of the surface and route of the old Mt. Hermon Road such that it was frequently closed in the wet winter months. After years of patching, repair and reconstruction by county road crews *(Figure 6.10)*, federal funds were ultimately obtained and the Mt. Hermon bypass was built, leaving the unstable hillside behind and allowing a safe and easy drive between Scotts Valley and Felton.

FIGURE 6.10. For many years the Mount Hermon Road was the most direct route between Scotts Valley and Felton. A large slump below the sand quarry continued to damage and distort the roadway until it was ultimately closed and the Mount Hermon Bypass was constructed.

The most landslide prone highway on the central coast, although not strictly speaking in the Monterey Bay area, is the Big Sur Highway. The coast from about Pt. Sur nearly to San Simeon was essentially isolated and not accessible by any vehicles until the highway was completed in 1937. Building this road was difficult from start to finish and keeping this highway open remains a yearly challenge. The mountains are among the steepest anywhere along California's coast and the rocks are notorious weak and prone to failure, particularly during very wet winters when the area may receive up to 80 inches of rain. The California Department of Transportation has named most of the large slides (Pitkin's Curve, Hurricane Point, Grayslip, Grandpa's Elbow, Big Slide, and Mud Creek, to name a few) simply because they have repeatedly had to remove the debris from rock slides and slumps in wet winters and in many cases, completely rebuild the highway. State Highway 1 has been closed for months at a time as the process of rock removal and reconstruction takes place. A 1983 landslide near Julia Pfeiffer Burns State Park led to the closure of the highway for over a year *(See Figure 6.1)*, with repair work costing over $17 million (in 2017 dollars). The wettest winter in recent California history (2016-17) again led to some very large rockslides, including the massive Mud Creek Slide, which closed the road for well over a year with repair costs reaching $57 million *(Figure 6.11)*.

After a landslide or slump has taken place, often during a very wet winter, the resulting landforms are easily recognizable. A scarp or low cliff and curved or concentric cracks typically mark the upper portion or head of the slide *(Figure 6.12)*. The main portion of a landslide is often distorted and broken up with cracks along the edges or margins, and the lower end or toe is of-

FIGURE 6.11. The Mud Creek landslide along the Big Sur coast in 2017 closed Highway 1 for over a year and pushed 40 acres of debris out into the nearshore waters. (Photo courtesy of Jon Warrick, United States Geological Survey).

ten seen as a bulge at the base of the hillside. A rotational slump, on the other hand, may have a surface that ends up relatively flat, and any trees present have probably been tilted, often uphill, sometimes downhill, but probably no longer standing vertical. These flattened areas along the middle sections of these large slumps on otherwise steep hillsides may look like perfect building sites to the innocent buyer or builder, and have often been used as such. On the positive side, an area which has already failed and readjusted to the forces of gravity is typically more stable than it was before; but unfortunately, adding the load of a house, putting excess water into the ground from a septic tank and leach field, or from rain gutter or driveway runoff, may lead to renewed instability and downhill movement.

FIGURE 6.12. This moderate size slump along Mar Monte Drive near Larkin Valley illustrates a set of common features: a head scarp, individual slump blocks separated by smaller scarps across the mid portion of the slump, lateral cracks and a bulging toe which moved onto Mar Monte Drive.

Perhaps the greatest geologic risk involved with building in the mountains of the Monterey Bay area is the potential to be affected by some type of future hillside or landslide movement. For many newcomers to the area, whether from the Santa Clara Valley, Southern California, or somewhere back east, the towering redwoods, the mountain air and seclusion are the only things they see. And even though realtors must provide, and buyers must now sign exhaustive disclosure statements regarding the presence of geologic hazards, very few realtors or purchasers are geologists, and many may be new to the area themselves. The number of mountain or hillside homes that were destroyed and damaged during the wet years of the last several decades from landslides and debris flows, or those damaged during the 1989 earthquake, are reasons enough for anyone desiring to buy mountain property to take a good hard look and get competent professional advice.

DEBRIS OR MUDFLOWS

Debris flows and mudflows are words used interchangeably and describe slope failure which is fluid rather than solid, and therefore flows downhill in contrast to the slower moving landslides just described. These flows usually begin within a depression or swale on a hillside where surface and/or subsurface runoff has accumulated to weaken the soils and weathered bedrock. Once these surface materials have become saturated with water they can liquefy and flow downhill along a fairly narrow corridor or path, perhaps only five to 25 feet wide, and then spread out in a fan-shape when they reach the base of the hillside or a flatter area. While mudflows are common in areas where the vegetation has been removed through forest or brush fires, or through logging, they also occur on undisturbed hillsides as normal events or processes.

Debris flows or mudflows form when several conditions are met. First, there must be an accumulation of several feet or more of permeable soil and weathered bedrock on top of more impermeable bedrock, and all of this on a relatively steep slope. Secondly, failure of this material takes place when there has been a long period of sustained precipitation, followed by high intensity rainfall that saturates the soils and weathered material resting on the bedrock, and converts it to a fluid or mud. The impermeable bedrock at depth stops the rainfall from penetrating further. This causes the water table to rise in the soils and changes its condition to that of mud or a viscous fluid. When this happens, the soils are no longer stable on moderate to steep slopes and they start to ooze and then flow downslope. If the slope is steep enough and the material is fluid enough, the mudflow begins to approach the consistency of floodwater and can move quite rapidly. With a flatter slope and less water the material will flow more slowly and probably not create as much damage.

Some mudflows may contain only 20-30 cubic yards of material, just a few dump truck loads, and take place on an unpopulated hillside in the middle of nowhere *(Figure 6.13)*. While it's an interesting event to a geologist, it has no significant impact other than modifying the hillside. At the other extreme, mud or debris flows may also contain hundreds or thousands of cubic yards of material, and can take place within or above populated areas, and be responsible for both property destruction and loss of life. These flows are potentially more dangerous than other types of mass downslope movements because they may move at speeds of up to 25 to 30 miles per hour – faster than people can run. They can cause considerable damage when they impact houses or other structures because of the velocity of the mud and debris and therefore its destructive impact.

FIGURE 6.13. This small debris flow occurred in the hills of Aptos during the prolonged rainfall of early January 1982. Rainfall and runoff were concentrated in a small swale or depression that served to liquefy the surface materials to the point of failure and downslope flow.

A forest and brush fire (the Molera Fire) swept quickly through the Big Sur area during the late summer of 1972, burning the vegetation over 4,000 acres of steep coastal mountains. These slopes usually receive 40 to 60 inches of rain annually. Short periods of very intense rainfall followed long saturating rain during the following November and led to extensive mud and debris flows on the steep burned over slopes. The debris flows, estimated at 10,000 cubic yards (1,000 dump truck loads) flowed down the steep slopes along the narrow drainages, closing Highway 1 and partially burying the picturesque village of Big Sur, including houses, businesses, automobiles, and mobile homes *(Figure 6.14)*. Boulders up to 10 feet in diameter and redwood trees up to three feet across were carried by the flows, and along with the mud, which hardened to the consistency of concrete, had major impacts on the structures and vehicles in the path of the flows *(Figure 6.15)*.

FIGURE 6.14. The Molera Fire in Big Sur during the summer of 1972 was followed by a wet winter that produced debris flows, which damaged and partially buried the Big Sur village.

FIGURE 6.15. The boulders carried by the 1972 Big Sur debris flows damaged or destroyed cars, buildings and mobile homes in the small village of Big Sur.

THE HISTORY OF SLOPE FAILURES IN THE MONTEREY BAY REGION

While we can extract large historic events such as earthquakes and floods out of the written record with some detective work, tracking down major episodes of landsliding and mudflows is considerably more difficult. In general, they have occurred as a result of either periods of prolonged or high intensity rainfall, or in some cases as a result of seismic shaking during large earthquakes. The two largest historic earthquakes in the Monterey Bay region, 1906 and 1989, both initiated widespread and well-documented examples of slope failure, from the sea cliffs to the summit of the Santa Cruz Mountains. Most of the other landslides and slope failures of the distant past are buried in accounts of individual storms and floods. The newspaper coverage of those long ago events is a function of how well traveled or populated the particular area was at the time, or perhaps what other competing events filled the front page. As a result, in general, we have treated the slope failures of the past several decades in more depth than those of earlier periods, simply because of the limitations of the available historic record.

HILLSIDE FAILURES PRIOR TO THE GREAT 1906 EARTHQUAKE

Shaking from an earthquake on October 8, 1865 (~7.0-7.3 magnitude) centered in the Santa Cruz Mountains brought down many boulders from the hillsides that obstructed the old Santa Cruz Gap Road. In addition, "*overhanging cliffs fell onto graded roads*" in the Corralitos area, and "*cliffs of rock and earth*" fell along the San Lorenzo River at the old Santa Cruz Powder Works (now Paradise Park).

The newspaper accounts of the storm and floods of the following winter (December 1866) describe numerous landslides throughout the Santa Cruz Mountains, especially along the road from Santa Cruz to Los Gatos, which was rendered impassable after washed out gullies, caved embankments and landslides.

Two years later, the October 21, 1868 shock (~7.0-7.3 magnitude), centered on the Hayward Fault in the East Bay, was credited with producing a 50-foot wide landslide along Majors Creek along the north coast, which carried rocks and trees 1,000 feet downslope. In February 1889, during a winter of heavy rainfall and flooding, numerous landslides occurred again on the mountain road leading to Los Gatos. So for those weary commuters of today who feel "Acts of God" are impacting their daily trip over the hill during the wet winter months, remember that the rocks were falling and earth was flowing onto the primitive mountain roads a century ago, and those poor folks were on horseback or in wagons. In 1890, an earthquake only described as "moderate," brought down landslides from the steep bluffs along Chittenden Pass, which closed both the highway and railroad.

SLOPE FAILURES FROM THE APRIL 18, 1906 EARTHQUAKE

The severe shaking produced by the 7.9 magnitude 1906 San Francisco earthquake induced the earliest well-documented major landslide events in the area. While most accounts of the earthquake focus on the urban areas, the most tragic and devastating events in Santa Cruz County, in terms of loss of human life, were the slope failures resulting from the earthquake. The April 18, 1906 quake followed a winter of way above average rainfall, which left the soils and hillsides weak and susceptible to failure.

What has often been overlooked in the accounts of this earthquake were the flooding and landslides initiated by the very high rainfall that preceded the April 18,1906 event. Boulder Creek, high in the San Lorenzo Valley, which typically has the highest rainfall totals in the Santa Cruz Mountains, had been deluged by 55.7 inches of rain during January, February and March, along with 16 inches in the previous four months. In the week between January 11 and 28, 1906, Wrights (just east of Highway 17 near Summit Road) was hit with 27 inches of rain. Landslides, rocks and mud blocked sections of the railroads and roadways at many locations throughout the Santa Cruz Mountains. Bridges were washed out from flooding in Soquel. "*The Loma Prieta Lumber Co.'s Mill at Hinckley Creek, 7 miles above Soquel, was swept away by the storm Thursday night (Jan.*

18) and hardly a trace of machinery, dwelling houses, barns, skid roads or wagon roads remain. Eleven men, including J. W. Walker, the foreman of the mill, camped all night Thursday on top of the mountain for safety and fortunately no one was injured by the wash out. Some of the debris was found on the beach near Capitola".

Three months later when the big earthquake occurred, very large landslides and debris flows that moved entire hillsides and forests downslope were common throughout the Santa Cruz Mountains. Although population densities in the hills were far lower than today, the locations of the particular hillsides that failed were the problem. The slopes were still very wet and the Loma Prieta Mill on Hinckley Creek, which had been seriously damaged in the January rainstorms, was struck with a devastating landslide. Nine men died as they slept in their bunkhouses when the mill was buried by a wet mass of earth 100 feet deep, while several hundred feet away others were spared. The speed of the landslide was described as "*extraordinary*" and dammed up the creek creating a lake 50 to 60 feet deep.

"*On Deer Creek* (a tributary of Bear Creek that drains the west slope of Castle Rock Ridge) *a large landslide started from near Grizzly Rock and slid westward, but changed its direction 60° or more farther toward the creek* (See Figure 2.20). *The shingle mill and houses in the creek bottom below the slide were buried under a reported 50 to 100 feet of earth, and two people were killed* (Figure 6.16). *It is 500 feet from the mill in the gulch to the top, at the point where the slide started. The slide covered about*

FIGURE 6.16. The 1906 San Francisco earthquake came after a winter of high rainfall in the Santa Cruz Mountains, which generated some large landslides. This slide along Deer Creek in the Santa Cruz Mountains partially buried a lumber mill and led to the death of one of the occupants (see figure 2.20). (Photo courtesy of the University of California Santa Cruz Special Collections).

25 acres of ground, and destroyed a lot of virgin timber from 3 to 10 feet in diameter. The slide material, which is 300 feet deep, is composed of soil, clay and shale. A witness watched as large redwoods on the slide mass performed 'all kinds of acrobatic feats.'

"On Bear Creek a smaller slide (than Deer Creek) *had moved a few hundred feet, buried a hut and killed one man". Opposite Boulder Creek another large slide was reported as having dammed the San Lorenzo River. Slides were also reported along Corralitos Creek, Henry Creek* (a short stream tributary to the west of Waddell Creek), *and at Alma* (a ghost town now buried beneath the waters of Lexington Reservoir). *Residents above Henry Creek stated that the earthquake "rolled boulders as big as a cottage into the creek and snapped off cottonwoods about 20 feet above the ground".*

In addition to the slope failures in the mountains, the seacliffs also responded to the severe shaking. In Capitola, it was reported that "*much earth fell from the bluffs near the town, but there was no appreciable effect on the surf...a continuous cloud of dust rose along the cliffs between Castro's Landing* (now called Rio Del Mar) *and Santa Cruz...*"

MORE RECENT SLOPE FAILURES

December 1955

The heavy rains that produced some of the most severe flooding in the county's history also led to widespread hillside failure. Highway 9 from Santa Cruz to Felton, as well as most of the other roads leading into the San Lorenzo Valley were blocked by trees and large slides of rock and earth. Seventy feet of the city's Laguna Creek pipeline was destroyed by a landslide at the point where the line crosses Smith Grade between the old Bald Mountain School and Bonny Doon.

April 3, 1958

Heavy rains took their toll on county roads as "*slides closed almost every road in the county at some time or other yesterday, but highway crews, working around the clock, had many reopened by today*".

January 22, 1967

Landslides can move more than rock and dirt. Following heavy rains in the San Lorenzo Valley "*a Huckleberry Woods residence...broke loose and slipped down a hill. Boulder Creek firemen and sheriff's deputies were summoned when it appeared that other structures were in danger. A couple of parked automobiles also slipped down hills at the location, which is just north of Boulder Creek ... Numerous slides were reported on such roads as Bonny Doon, Lompico, Bean Creek and Creek Drive.*"

January 3-5, 1982

Late in the afternoon of Sunday, January 3, 1982, a light rain began to fall in the upper reaches of the San Lorenzo Valley. The rainstorm, coming out of the southwest, rapidly increased to monsoon intensities and sustained that level for about 30 hours until late Monday, clearing just before midnight. The small mountain communities of Boulder Creek, Ben Lomond and Lompico represented "*ground zero*" for this particular storm. Rainfall in these areas totaled 12 to over 15 inches during the 30-hour storm period; about one-third of the average annual precipitation fell in just over a day.

Many residents who left for work Monday morning were unable to return home that evening as much of the San Lorenzo Valley was cut off by either swollen river and creek channels or the landslides, mud and fallen trees that covered many of the mountain roads. North of Ben Lomond along Love Creek, the stranded residents of Love Creek Heights were set to weather the storm. The almost 40 homes scattered across the hillside above Love Creek ranged from older summer cabins to modern homes. During the evening hours of January 4th, they kept busy trying to stay dry and warm and maintaining the drainage systems and culverts around their homes.

The storm ended abruptly around midnight on January 4th and the sky was filled with stars. The only indication that a storm had recently deluged the area was the roar of Love Creek at the bottom of the canyon, raging out of its banks and tearing out the roadway. In the very early hours of Tuesday morning, most of Love Creek's residents were asleep. And then at about 1:00 a.m., a 1,000-foot long slab of the steep and rain-weakened hillside above them broke loose, and rapidly swept more than 600,000 cubic yards of rock, mud and debris down through Love Creek Heights towards Love Creek *(Figure 6.17)*. This may have been one of the largest and most devastating landslides to ever impact the Santa Cruz Mountains and the central coast. While the Big Sur area has been repeatedly hit by larger slides (the Mud Creek Slide of May 2017, for example, involved an estimated 2.4 million yds^3), rarely have lives been lost.

FIGURE 6.17. The massive Love Creek Slide above Ben Lomond in early January 1982 involved 600,000 cubic yards of rock and soil and buried nine homes and ten people.

In its path, this 35-foot thick flow of rock and earth buried nine houses and ten people. Sadly, some never woke again. Most of those who died were apparently caught asleep in their beds, attesting to the rapid rate of the slide's descent into the canyon. Four others barely escaped with their lives and were rescued from the mass of landslide rubble and splintered homes the next day *(Figure 6.18)*. Search parties were later to recover six bodies as they dug into the mud at the base of the landslide, but four persons were never found.

FIGURE 6.18. One of the homes destroyed by the Love Creek Slide in 1982.

One of the survivors of Love Creek, Karen Wallingford, a 32-year old schoolteacher, wanted to die as she lay beneath the wreckage of her home on the morning of January 5th. With her legs and head pinned by debris, she thought about her family in Massachusetts. She wasn't afraid but wished she would die. As Karen was sleeping in her glassed-in porch, which served as her bedroom, she heard trees crashing down outside in the darkness. Knowing it was close, she got out of bed quickly. In an instant, a wall of mud destroyed the living room. She remembered being flung against the wall as furniture was thrown towards her. When everything stopped moving, she was trapped beneath the wreckage with only a tiny space in front of her. She was able to move only the fingers on one hand and thought only her room had been damaged. When her calls got no response from her two roommates, she realized the extent of the damage. Shortly after rescuers found her pinned beneath the rubble the next morning, her friends' bodies were discovered buried in the wreckage of their bedroom.

On the basis of loss of life, this was the most destructive, naturally induced landslide at the time in California's history, and the third most tragic in western North America in this century. This storm, which struck the entire San Francisco Bay area, induced more than 18,000 debris flows, which swept down hillslopes or drainages without warning, damaging at least 100 homes and killing 14 Bay area residents. In Santa Cruz County, 22 died. Love Creek was the worst of the lot, and the largest of thousand of landslides and debris flows in the Santa Cruz Mountains *(Figure 6.19)*.

FIGURE 6.19. A house along Love Creek was half buried by the mud brought down during the 1982 slide into the creek bottom.

Between 7:30 and 8:00 p.m. on January 4th, a slump from above Alba Road, near Ben Lomond, liquefied into a debris flow that flowed 650 feet to Highway 9. This flow totally destroyed one house and four cabins, and substantially damaged another house, two other cabins, a preschool, and two utility buildings.

In northeast Scotts Valley off Canham Road, several small slides on steep slopes above a small valley mobilized as debris flows that crossed an old apple orchard and coalesced into a narrow canyon. Two boys walking along a road in the canyon were picked up by the flow; one boy was thrown clear where he grabbed a large tree as his clothes were torn off by the torrent, while the other was carried away and drowned in Carbonera Creek.

Another similar failure was initiated below Lake Boulevard in Lompico and traveled rapidly down a long, narrow drainage, crossed Creekwood Drive, and flowed into Lompico Creek, a distance of about 650 feet. The flow destroyed or damaged several houses in its path by toppling trees onto them, battering them with rocks and mud or piling debris against them until they collapsed. One eyewitness described the flow as initially moving as fast as a frightened man can run at full speed downhill. The material in the flow had a consistency of stiff, wet concrete.

FIGURE 6.20. The intense rainfall of early January 1982 produced slope failure from the summit of the Santa Cruz Mountains to the seacliffs. Along Beach Drive in Rio Del Mar a debris flow from the coastal bluff pushed this house off its foundation and out onto Beach Drive.

In addition to the dozens of slope failures in the Santa Cruz Mountains, the seacliffs suffered from the impact of over eight inches of rain in 30 hours. Damage was greatest along Beach Drive in Rio Del Mar as the steep, rain-weakened bluffs failed, damaging houses built on the beach below. The bluff top failure threatened a number of houses as foundation support was removed, and destroyed several beach homes due to debris impact *(Figure 6.20)*. As is often the case, homeowners at both the bottom and top of the bluff filed lawsuits blaming the damage on the other party.

The combined damage from the 1982 storm-related flooding and landsliding led to a total of 102 lawsuits totaling in excess of $147 million (in 2017 dollars) filed against Santa Cruz County, and 51 suits claiming $127 million (in 2017 dollars) filed against the city of Santa Cruz. For damages suffered from many geologic disasters, which may not be covered by homeowner's insurance, the frequent response today seems to be to find someone to sue for any damages, typically a government entity, as often encouraged by newspaper ads posted by law firms. What ever happened to "acts of God?" Unfortunately, as we all know, the deep pockets don't exist any longer. Most local governments are self-insured today and have to use the same funds that pay for roads and schools, police and fire, and water and sewer, to defend themselves against the countless annual lawsuits resulting from excess rainfall, landslides or random earthquakes. Most homeowners' policies insure houses from fires and a few other hazards, but don't insure against the natural disasters right outside the front door.

The El Niño Winter of 1983

While the winter of 1983 was best known for its coastal storm damage (~ $255 million in 2017 dollars along the entire length of the state's 1,100 mile coastline), the same storms that brought destructive waves to our cliffs and beaches, also brought rain clouds and heavy precipitation to the Monterey Bay area. Boulder Creek and Lompico received over 100 inches of rain, or over eight feet, that winter.

One of the areas most affected was the small mountain community of Lompico. An ancient landslide was activated by the long saturating rainfall and in mid-March, an estimated 100,000 to 200,000 cubic yards of earth and mud (10,000 to 20,000 dump truck loads) oozed slowly across Lompico Road isolating nearly 2,000 residents. As road crews attempted to move the mud at the toe of the slide along the road, more material moved in to replace it. The solution proposed and ultimately carried out, was to bridge Lompico Creek and build a 1,000 foot long bypass road along an old logging trail on the opposite side of the creek, and then cross the creek with another bridge.

The same March storms and accumulation of rainfall closed Highway 9 above Boulder Creek with a large mudslide, and also isolated an area of Lockhart Gulch with a massive slide that caught a car as it moved across the road. Highway 1 in the Big Sur area was closed for months by a series of large slides including a very large failure at Sycamore Draw. CalTrans ultimately spent over $17 million (2017 dollars) to reshape the entire mountainside and push the material over the cliff into the sea *(See Figure 6.1)*. The closure of 60 miles of the scenic highway wreaked economic havoc on the businesses that rely almost exclusively on the tourists who pass through the area.

The Winter of 1986

For the third time in five years, well above average rainfall soaked the hillsides of the mountains surrounding Monterey Bay and weakened them to the point of failure. The Valentine's Day storm in mid-February was one of the most intense periods of rainfall throughout the entire winter. During the week of February 13-20, Ben Lomond received 26.10 inches of rain, Boulder Creek 26.05 inches, and Empire Grade 23.51 inches. Newspaper accounts on February 18 listed the following roads closed due to landslides, mudslides, or down trees:

San Lorenzo Valley: Highway 9 five miles north of Santa Cruz, Graham Hill Road, Lompico Road, Glenwood Drive, Newell Creek Road, Two Bar Road, Lockhart Gulch Road, Hubbard Gulch Road, East Zayante Road, Boulder Brook Drive and Redwood Lodge Road.

Central Santa Cruz County: Brookwood Road, Glen Canyon Road, Redwood Lodge Road, Schulties Road, Morrell Cutoff, Laurel Road, Granite Creek, Branciforte Drive, and Porter Street in Soquel were all closed.

One woman was killed in the Boulder Brook area off Big Basin Way as a large debris flow (100 feet long, 150 feet wide, and about 25 feet deep) moved rapidly down a steep hillside into Foreman Creek and swept her small wood frame cabin off its foundation and deposited it 100 feet downstream against a stand of redwood trees. A number of the Boulder Brook residents were evacuated because of fears of additional debris coming off the water-soaked slopes.

Three homes in the Hubbard Gulch area were threatened, one damaged, by a large debris flow that cleared an entire hillside and dammed Marshall Creek. A number of large redwood trees were caught up in the slide and were strewn across the creek creating a large dam of mud and debris.

The south county also received heavy rain, in particular the Eureka Canyon area, which was hit by a combination of mudslides and falling trees as well as floodwaters *(Figures 6.21 and 6.22)*. A number of cabins were demolished and two women were seriously injured in the steep canyon bottom as large trees crashed through their homes. Both waited long hours for rescue as California Department of Forestry workers cut their way with chainsaws and fought through rocks, mud and waist-deep water to reach them. The same flow of mud ripped a bedroom off a house about twenty feet away. One eyewitness said the mud flowed faster than he could run.

FIGURE 6.21. The 1986 Eureka Canyon slides and floods destroyed parts of the access road and isolated residents.

Traffic was at a standstill on Highway 9 just north of Brookdale as half of the roadway collapsed into the San Lorenzo River, leaving a steep muddy hillside plunging down to the river below with plastic utility pipes sticking out in every direction. Telephone crews trying to establish contact between Boulder Creek and rest of the world complicated the job of local firemen who were directing traffic around the slide along a single lane. Highway 9 is a delicate umbilical cord through the San Lorenzo Valley and when it goes out, vehicular and communication to the outside world goes with it.

October 17, 1989 Loma Prieta Earthquake

The 6.9 magnitude Loma Prieta earthquake produced widespread slope failure in the Santa Cruz Mountains and along the seacliffs of northern Monterey Bay due to the intense seismic shaking. While geologists who had worked in the Santa Cruz Mountains had interpreted much of the anomalous and hummocky topography along the summit and flanks of the mountains as being due to ancient landslides, formed either during pre-historic earthquakes or very wet winters *(See Figure 2.6)*, or perhaps a combination of the two, there was no easy way to test this conclusion. During the fifteen seconds of strong shaking in the early evening of October 17, however, the hillsides were tested.

A number of steep ridge top areas along the summit area were extensively cracked during the earthquake *(Figure 6.23)*. This ground cracking dam-

FIGURE 6.22. Heavy rainfall in the winter of 1986 led to widespread landslides, debris flows and flooding in the narrow Eureka Canyon above Corralitos.

aged dozens of homes as well as State Highway 17 and a number of county and private roads. While the origins of the cracks were complex and no single explanation could account for all of them, the conclusions of the geologists who worked in the area following the earthquake were that the majority of these features were apparently related to old landslides, mobilized by the intense ground shaking. The earthquake occurred in a dry autumn following two years of drought. Had the earthquake taken place during the winter or spring following a wet winter, as did the great San Francisco earthquake in 1906, landslides and mudflows would have been more widespread and would have caused substantially more destruction.

FIGURE 6.23. A number of very large dormant landslides in the Summit area of the Santa Cruz Mountains were briefly reactivated by the Loma Prieta earthquake, but because of the very dry conditions in October of 1989, extensive cracking developed but motion was very limited.

Many residents of the mountains, particularly the Summit Ridge area, which is bisected by the San Andreas Fault, were caught off guard by the cracks from the reactivated landslides, and were not convinced of their origin. The historic record is quite clear, however. Earthquake-induced landslides have been documented for at least 2,000 years, and have caused tens of thousands of deaths and billions of dollars in economic losses during the present century alone.

FIGURE 6.24. Reactivated scarp along an old landslide in the Summit area during the 1989 earthquake led to serious damage to this driveway and house.

The U. S. Geological Survey in a regional reconnaissance determined that landslides generated by the Loma Prieta earthquake occurred throughout an area of 5,000 square miles with the highest concentration in the Santa Cruz Mountains. Areas previously identified on county maps as ancient landslides in the Redwood Lodge Road, Laurel Road, Schulties Road, Morrell Cutoff areas, as well as the Villa del Monte subdivision were among the areas hit hardest by the reactivation of these large old landslides. In most cases, the effects on the ground surface consisted of deep cracks, low (up to three feet) scarps or ledges formed where the slide mass slipped downslope a few feet. These cracks usually formed in an arcuate or curved pattern, facing downhill, and outlined the margins of older, flat landslide areas. In many cases, particularly in the Villa del Monte area, these cracks went directly through homes, driveways and even sheared off water wells in the subsurface (*Figures 6.24 and 6.25*). Even a foot or two of movement, however, combined with the earthquake shaking, was enough to either seriously damage or destroy many homes.

FIGURE 6.25. Headscarp cracks on an ancient landslide resulting from the Loma Prieta earthquake in the Villa del Monte area of the Santa Cruz Mountains.

One concern, based on the landslides both during the 1906 earthquake and the continued sliding during the subsequent wet winter of 1907, was that rains in the winter of 1989-90, following the Loma Prieta event, would likely further weaken the hillsides and produce additional movement and damage. Fortunately for the homeowners, another year of below average rainfall failed to produce any significant movement along the cracks that had opened during the earthquake.

Some Summit area residences were damaged beyond repair and were demolished; some were rebuilt or repaired, and others sat for years, victims of the earthquake, and the uncertainty as to the risks posed by the cracks. The risk issue was and will always be a difficult one for local government officials, geologists, engineers, and insurance companies to deal with. Many residents wanted to return to their homes, fix them up and get on with their lives, but with some assurance that this wouldn't happen again. There are few people in any of the above professions who are able or willing to give these assurances, however, which made future damage and liability a sticky and unresolved issue.

Because of the short distance from the epicenter of the 1989 earthquake to the coastal cliffs and bluffs, shaking was intense in these areas as well. In fact, slope failure along coastal cliffs extended from Marin County, 70 miles north of the epicenter, to Big Sur, 80 miles to the south. Most of the cliff failures, however, took place within five to ten miles of ground zero in the Forest of Nisene Marks above Aptos Village. Between Opal Cliffs and Sunset Beach, cliff failure was widespread *(See Figures 2.37 and 6.6)* and damaged a number of bluff top homes. In addition to the direct fall of rocks and earth to the beach, fissures through

FIGURE 6.26. Sliding along this coastal bluff above Place de Mer in the Manresa Beach area during the Loma Prieta earthquake partially blocked the access road.

FIGURE 6.27. Slope failure above Place de Mer in 1989 overtopped this retaining wall and buried a car at the base of the slope. The old tires were ineffective in stabilizing the slope.

the weakened bluff top soils extended as far as 30 feet inland, cracking streets, patios and retaining walls, as well as foundations and walls of homes.

Perhaps the best example of this problem is along Depot Hill in Capitola, every geologist's favorite spot to go look at the effects of coastal erosion on structures and streets. The Crest Apartments on Grand Avenue were constructed in 1967 at the edge of the cliff with very minimal setback. Continued erosion and undercutting of the bluff over the years led to several attempts to shore up the foundation support system for the apartments. Cracking and failure of the cliff during the earthquake, however, undermined the foundation further, cracked the slab and perimeter walls of several units, and also removed support for several of the concrete caissons that were supporting the structures *(See Figure 2.38)*. Although a cliff protection scheme was being planned at the time of the earthquake, the loss of bluff material led to demolition of six damaged units and the closure of much of Grand Avenue to through traffic.

Nearly continuous sliding of the bluffs took place from Pot Belly Beach (adjacent to New

Brighton Beach State Park), through Las Olas Drive *(See Figure 6.7)* and Seacliff State Beach, to Rio del Mar's Beach Drive and Aptos Seascape *(See Figure 6.6)*, to Manresa State Beach and Place de Mer *(See Figure 2.37)*. Slabs of the sandy bluffs several feet to perhaps ten feet thick were shaken loose and slid to the base of the bluff where they collected as piles of sand and loose debris. These failures again cracked foundations at the bluff top, leading to destruction of at least three homes, with the debris accumulated against or on roads blocking access to some homes at the base of the bluffs and burying cars *(Figures 6.26 and 6.27)*. Had the duration of shaking been longer and rainfall been higher prior to the earthquake, coastal bluff failure would have been even more extensive and damage far greater.

While earthquakes of this magnitude do not occur frequently, there is a lesson to be learned for ocean front residents. Protecting that part of your home exposed to ocean waves is important, but homeowners also have to be aware that for those choosing to live at the base of a coastal bluff, there are other terrestrial forces to contend with as well. You can't turn your back on either the ocean in front of you or the bluff behind you.

SOME FINAL THOUGHTS ON LANDSLIDES AND OTHER SLOPE FAILURES

In a geologically young and active area like the Santa Cruz Mountains or Big Sur, the weak nature of many of underlying rocks, the steep slopes, whether along the coast or higher in the mountains, the high potential for prolonged and intense rainfall, and the seismically active nature of the region, combine to produce a landslide or debris flow-prone landscape. Adding to the natural factors creating instability are the many human activities that have sculpted and disturbed the natural terrain in the Monterey Bay area: grading for highways, roads, and building sites; excess water from septic tanks and runoff from impermeable surfaces, to name a few. Slope failures have occurred repeatedly in the past and can be expected to occur in the future.

The accounts of many of the older hillside failures were well-documented in the historic newspaper articles, but aren't always evident in the landscape itself; some are and some are not. The vegetation recovers and regrows quickly in places like the Santa Cruz Mountains, camouflaging the hillsides and often leaving little to indicate that a debris flow had occurred, destroying a home 25 years earlier. This presents challenges to both geologists who are often hired to perform site inspections, and also potential homeowners who, in the middle of summer, have no idea of what could happen in the middle of a very wet winter.

Chapter 7

Droughts

I want to know... Have you ever seen the rain... Comin' down...
– John Fogerty

INTRODUCTION

Thinking on the bright side, with our glasses half full – or perhaps overflowing – living here along the coastline of Monterey Bay, we don't have to deal with hurricanes, tornadoes or volcanoes. There is no shortage of natural disasters to contend with, however, but compared to the calamities discussed in earlier chapters – earthquakes, floods, debris flows and El Niños – droughts are subtle, and much more difficult to research, document and quantify. Unlike an earthquake, a drought is not instantaneous, and the effects are spread out over a much longer period of time. They are a gradual phenomenon and the impacts are usually felt first by those most dependent on annual rainfall – ranchers engaged in dry land livestock grazing, farmers dependent on irrigation or imported water, and rural residents relying on wells in low-yield aquifers or small water systems lacking a reliable water source. Drought impacts will increase with the duration of a drought, as reservoirs levels drop, and groundwater levels in aquifers decline.

California now has about 150 years of rainfall history, and throughout much of the last century, Chambers of Commerce and the real estate profession often claimed that the state had the greatest climate in the world. When severe floods occurred or the rains missed California for months on end, these years were often described as "*exceptional*" or "*unusual*". Before the construction of large reservoirs, Central and Southern California were usually equally subjected to floods and droughts. Some of the oldest newspaper accounts documented the floods of 1815-25, followed by a drought from 1827-29, and then the floods of 1832 and 1842.

The 1856-57 drought followed 20 years of near normal or above normal precipitation, which led to overstocking the luxuriant pastures. During the summer of that year and the subsequent winter, an estimated 100,000 cattle were lost in Los Angeles County alone. Following the legendary floods of 1861-62 came the drought of 1862-64, which may have been the driest of all historic droughts in Southern California. As will be described later, these dry years more or less brought about the end of beef cattle as the state's primary industry.

What constitutes a drought? In some cases we can't even agree on this. One simple definition of drought is "*a prolonged period of abnormally low rainfall*". And a new term has been added to our weather conversations in recent years, *mega-drought*, defined as "*a prolonged drought lasting two decades or longer*". This is a word most water suppliers would rather not think about, but evidence for such events is clear in the pre-historic record of tree rings.

When does a drought end? For the state as a whole, about one-third of our water comes from the Sierra snowpack, another third from groundwater, and rest from surface runoff, much of which is captured in reservoirs. So after a year or two or more of below average precipitation, we need enough rain and snow falling in the right places to bring the snowpack up, refill the reservoirs and recharge the aquifers – which can take several years or longer – to end a drought.

Santa Cruz County is one of only a handful of counties in California that is completely independent of outside water sources (other than those millions of bottles of Evian, Pellegrino, Fiji, Perrier, Aquafina, and Dasani they get trucked in). So while much of California depends on the statewide system of dams, reservoirs, canals and pipelines that was built over the past century or so to store winter runoff and move the water around to where it is needed, or where politics decides to pump it, Santa Cruz County has no external water source. When we have a dry year or two, we feel it. There are a number of different water providers in the county that spread out the responsibility for making sure we have water when we turn on the tap. There are, however, really only two water sources at present: surface runoff or groundwater, and ultimately both depend on water falling from the sky. And despite an old legend that has persisted for decades, the water we see in the Monterey Bay area doesn't get here through underground passages from the Sierra Nevada.

Monterey County is much larger (3,771 square miles in contrast to 445 square miles for Santa Cruz County) and its water supply is a bit more complex. In part this is because the northeastern or inland border of Santa Cruz County runs right along the crest of the Santa Cruz Mountains, so if the rain doesn't fall on the southwestern side of the mountain crest, we don't see it. Monterey County, on the other hand, extends through the Coast Ranges and southward nearly 100 miles down the Salinas River valley. There are two large reservoirs on Salinas River tributaries, Nacimiento and San Antonio, which collect runoff from tributaries in the Santa Lucias. The vast agricultural industry along the Salinas River relies on water from these two reservoirs and also groundwater from wells, which spreads out the water over a longer time period than surface runoff captured in the reservoirs. The Monterey Peninsula, on the other hand, being underlain by granite – which generally doesn't store large volumes of water – has little water of its own. The reservoirs behind the former San Clemente Dam and Los Padres Dam in the Carmel Valley were the major water sources for the peninsula cities for decades. They are relatively small, however, and rainfall in the Carmel Valley is generally quite low, averaging just 19 inches a year, so they are very drought sensitive.

San Clemente Dam was demolished in 2015 because of seismic safety concerns and the loss of nearly all of its storage capacity due to sediment fill *(Figures 7.1 and 7.2)*. The reservoir behind Los Padres Dam is the last water storage facility in this area, and is the primary source of water for Carmel Valley and the Monterey peninsula. Water from the Los Padres Reservoir is also released to recharge the Carmel Valley aquifer. Discussions have been underway for some time on how to deal with Los Padres Dam and what other options exist for providing water for the Monterey Peninsula, desalination for example.

There are difficulties in describing and comparing historic droughts with modern dry spells such as the 2012-2016 period that the state broke out of – at least temporarily – in 2016-17 *(Figure 7.3)*. The demographic, population and economic conditions are quite different today than they were 50, 100 or 150 years ago. California's population has grown from 380,000 in 1860, to 1.5 million in 1900, 10.6 million in 1950, 20 million in 1970, 30 million 1990 and is approaching 40 million in 2018. The economic base has shifted on the agricultural side from cattle grazing, which dominated until the latter part of the 1800s, to California becoming the largest producer of fresh fruits and vegetables in the nation. The growing population with increasing demands for water, as well as the development of a water intensive agricultural base – in contrast to cattle grazing on open ranges – are just two examples of changing water demands. Several dry years that may not have been considered a drought a century ago, when the state had just three million people, present vastly different challenges today with a population of nearly 40 million and 7.5 million acres of crops dependent on regular irrigation.

FIGURE 7.1. San Clemente Dam in the Carmel Valley before removal showing the sediment that had nearly completely filled the reservoir behind the dam. (Photo courtesy of California American Water (CalAm)).

FIGURE 7.2. The San Clemente Dam during removal. (Photo courtesy of California American Water (CalAm)).

HISTORIC DROUGHTS AND MEGADROUGHTS

Not surprising to long-time California residents, droughts are not new or uncommon. We get these dry periods every decade or two. If you really want to get concerned and lose sleep, the pre-historic record has preserved far more serious droughts in the distant past than we have seen over the past 150 or so years of record keeping in California; megadroughts in today's terminology. But this was before California was home to 40 million people and our farms and fields were providing fruits, vegetables and livestock to the entire nation. In the early years of the last century, if there wasn't enough water where we wanted a city or farm, we built dams, canals and pipelines to move the state's water to where political pressure demanded it. And more often than not, we fought over who had the rights to the water. In the "*Owens Valley Water Wars*" of 1924 and 1927, farmers protesting Los Angeles stealing their Owens Valley water actually dynamited aqueducts. Nonetheless, between 1860 and 2000, 1,400 larger dams (over 25 feet high) were built on California's rivers and streams; that's an average of 10 per year or almost one every month for 140 years.

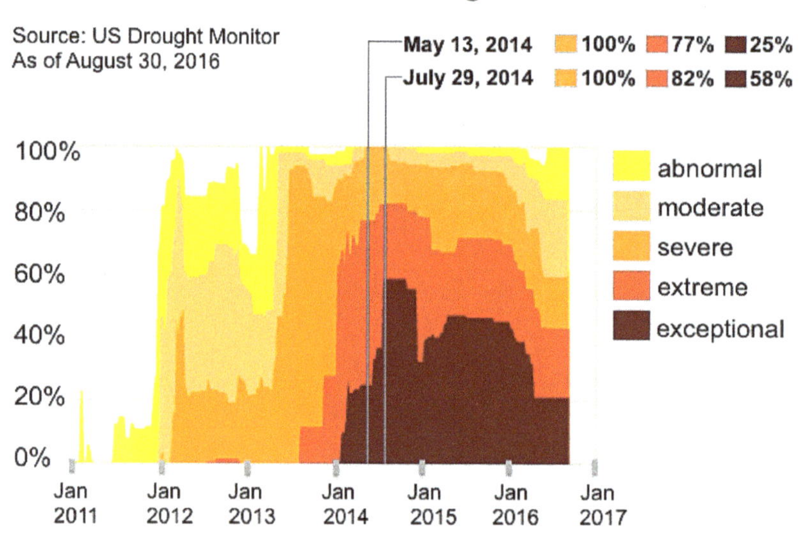

FIGURE 7.3. The percentage of California in various drought conditions throughout the 2012-2016 drought. (Illustration courtesy of the United States Drought Monitor).

Droughts over the past two centuries in California have usually lasted several years, and there are many accounts of how the dry periods in the 1800s affected livestock, agriculture and those living here at the time. Until the last decade or so of the 1800s, however, there were no reservoirs to store winter runoff, so low rainfall years hit the early residents particularly hard. An important question to try and answer as we look towards the future is whether the last 200 years of climate and rainfall have been typical.

Rainfall records in the state only go back about 150 years, but dendrochronology, or the study of tree rings, has allowed us to look much further back in time. Just like putting on a few extra pounds around your waist when you eat well on a vacation, trees suck up moisture when there is plenty to go around, and use that extra water to grow thicker growth rings.

History has been written and recorded in many places other than in books, and the job of a paleoclimatologist is to find where those climate records have been preserved. Tree rings, lake and seafloor sediments, corals, and ice are a few places where we have been successful in extracting long-term climate records.

Bristlecone pines, which can live to be 4,000-5,000 years old, and survive in the White Mountains of southeastern California, are living history books in which the records of our prehistoric rainfall have been preserved. These ancient trees contain the evidence in the

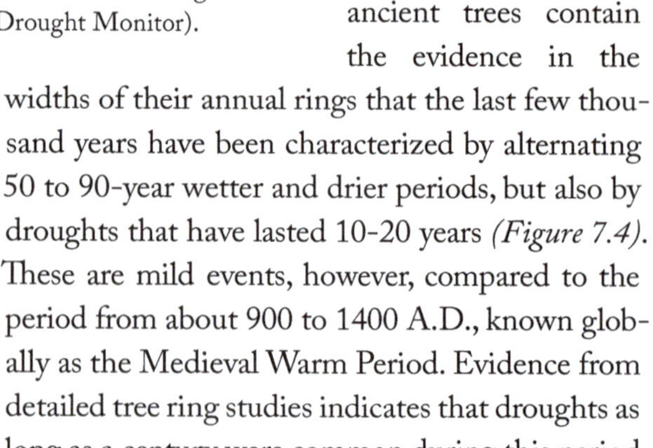

widths of their annual rings that the last few thousand years have been characterized by alternating 50 to 90-year wetter and drier periods, but also by droughts that have lasted 10-20 years *(Figure 7.4)*. These are mild events, however, compared to the period from about 900 to 1400 A.D., known globally as the Medieval Warm Period. Evidence from detailed tree ring studies indicates that droughts as long as a century were common during this period.

Most climate scientists would agree that the past century was unusually wet, and it has been

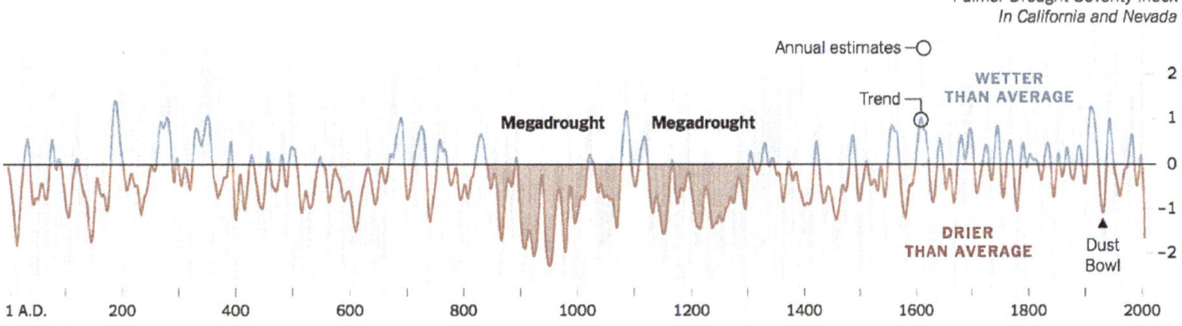

FIGURE 7.4. The long-term drought history of the Western United States, including extended dry periods known as megadroughts. (Image courtesy of the *New York Times*).

during the past 100 years that the populations of the arid southwestern states literally exploded. Nevada went from 42,000 people in 1900 to just over three million in 2018. Arizona grew from 123,000 to about seven million during the same period. And these states are both deserts with a lot of thirsty golf courses.

California has nearly 40 million people today, over 25 times more than in 1900. And all this growth, whether farms, factories or cities, took place during what was very likely an unusually wet century. For decades, Central Valley agriculture had everything – sunshine, fertile soils, and water. But that water was imported from elsewhere or pumped from aquifers that were being depleted year after year. In 2015, the fourth year of drought, 560,000 acres of productive farmland in California were left fallow due to the lack of available water. As of late August of that year, 95% of the entire state was designated as being under severe drought conditions, and most of the largest reservoirs were at just 25–33% of their total capacity.

If we are again entering a longer-term drought, the impacts and implications for California are enormous. It's hard for many to believe this is possible, but history tells us it has happened in the past. We have lived our entire lives in times of abundant water, but we need to begin coming to grips with the potential of a different future.

THE DROUGHT OF 1862-1864

There are numerous records or accounts of the drought history of the American Southwest, which list specific dry periods and their impacts *(See Figure 7.4)*. There seems to be some general agreement among these sources that the drought of 1862-64 had some of the greatest impacts of any in the 19th and 20th centuries. Until this severe dry period, a major part of the young state's economy and even way of life, primarily that of the Californios, was tied to cattle. Californio is a Spanish term generally defined as "*persons of Spanish or Mexican heritage whose place of birth or residence was California*". In most cases these were the early residents who were given the major Mexican land grants, many of which became the large cattle ranches. And many geographic locations in California were named after the original Californios – Pico, Vallejo, Alvarado, Castro, Pacheco, Carrillo, and many others.

In the words of local historian, Sandy Lydon, "*the drought of 1862-64 was incredibly hard on the Californios who were trying to hang on to their pas-*

toral culture. Many California historians have written that the 1862-1864 drought ended the Californio culture forever, and that is certainly the case in Santa Cruz County".

"Strapped for cash, the Californios went to the money-lenders and put the only thing they owned – their land – up for collateral. It had never stayed dry for longer than one season, they reasoned. Surely it would rain in December, or January. But the storms stayed north and the next spring and summer the sound of the auctioneer's hammer echoed above the bawling of the dying livestock. Some ranchos fell for 10 cents an acre. David Jacks bought the 8,794 Los Coches Rancho (Monterey County) for $3,535. The names Garcia, Soberanes and Castro were removed from deeds and Spence, Jacks, Iverson and Hihn put in their place."

In an excerpt from "*Exceptional Years: A History of California Floods and Droughts,*" J.M. Guinn in 1890 wrote that rainfall in "*1862-63 did not exceed four inches, and that of 1863-64 was even less* (this appears to have been describing southern California). *In the fall of 1863 a few showers fell, but not enough to start the grass. No more fell until March. The cattle were dying of starvation… The loss of cattle was fearful. The plains were strewn with their carcasses. In marshy places and around the cienegas* (loosely translated as wet meadows), *where there was a vestige of green, the ground was covered with their skeletons, and the traveler for years afterward was often startled by coming suddenly on a veritable Golgotha – a place of skulls – the long horns standing out in defiant attitude, as if protecting fleshless bones.*"

THE EXTENDED DROUGHT OF 1929-1934

The 1929-1934 drought occurred during the infamous Dust Bowl period that impacted the Great Plains of the United States in the 1920s and 1930s and brought many new immigrants to California. It occurred within the context of a decades-plus dry period in the 1920s and 1930s *(Figure 7.5)*, whose hydrology rivaled that of the most severe dry periods in more than a thousand years of reconstructed Central Valley paleoclimatic data. The drought's impacts were small by present-day standards, however, since the state's urban and agricultural development and therefore water demands were far less than those of the 21st century. The state's population in 1930 was just 5.6 million compared to nearly 40 million in 2018. In response to this drought, however, the vast Central Valley Project to move water around California – from where it was to where it wasn't – was begun in the 1930s.

Locally, the city of Santa Cruz recorded seven consecutive years of below normal precipitation between 1928 and 1934 (22.1", 17.85", 21.47", 12.47", 27.3", 21.65", and 18.25", respectively *(Figure 7.6)*, where the long-term average is 30 inches. Unfortunately, neither Monterey nor Salinas have official rainfall records for this entire period, although Salinas did record just a little over 10 inches in both 1932 and 1933, and 9.6 inches in 1934, all somewhat below the long-term average of 13 inches. While this was the most severe drought in decades, the populations of both counties were quite small at the time. Santa Cruz County had just 37,433 people in 1930 while Monterey County had 53,705. By the time of the next major drought in 1976-77, Santa Cruz County had over four times as many people to provide water for (~160,000), and Monterey County's population had increased five fold to about 275,000. While the numbers of people in both of these counties have continued to grow, the available water supplies haven't increased, so in dry years, the shortages or deficits continue to get larger.

THE DROUGHT OF 1976-1977

While short in duration, this was one of California's and the Monterey Bay area's driest two-year periods on record *(See Figures 7.5 and 7.6)* and served to wake up many of California's water

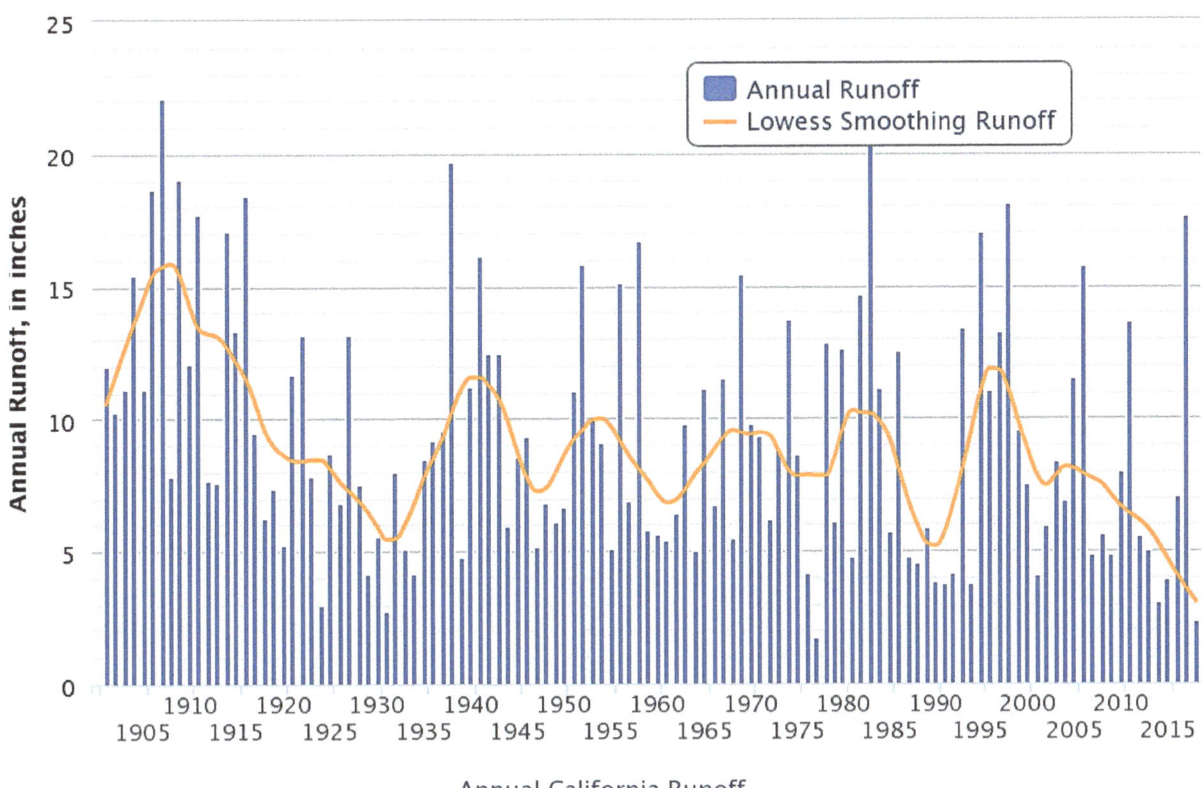

FIGURE 7.5. Annual runoff for the entire state of California from 1900 to 2017. Annual runoff is a measure of the average amount of water draining off of California in inches by year. (Graph courtesy of the United States Geological Survey).

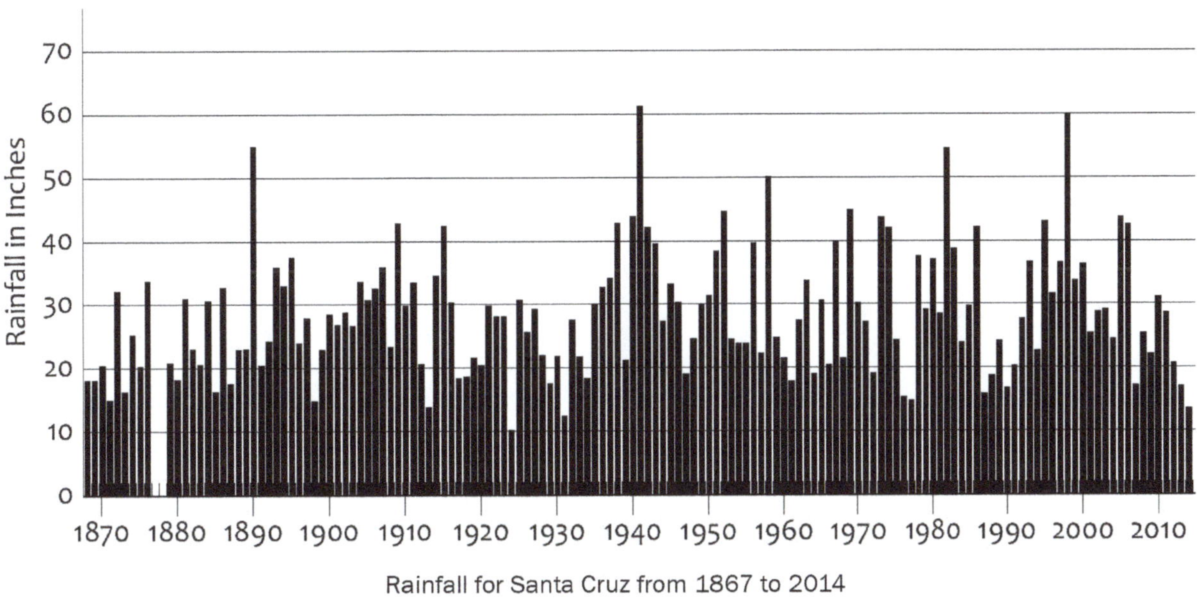

FIGURE 7.6. Annual rainfall for Santa Cruz from 1867 to 2014. Average is ~30 inches. Gap in 1877-1878 is due to lack of data rather than complete lack of precipitation. (Graph based on rainfall data from the Santa Cruz City Water Department).

agencies that they weren't prepared for major reductions in their supplies. Forty-seven of California's 58 counties declared local drought-related emergencies with mandatory water rationing. Precipitation for most parts of the state was less than half of normal, with some places receiving just 15 percent of their long-term average. River flows declined precipitously and reservoirs levels dropped dramatically in response.

With a much larger population than during the previous drought, the city of Santa Cruz for the first time implemented water use restrictions, also known as water rationing, which is never popular with anyone. The city recorded just 13.88" in 1976 and 15.93" in 1977, the two-year total being equivalent to the average rainfall for a single year. Monterey (average annual precipitation of 19.31") had comparably low rainfall, with just 9.76" falling in 1976 and 10.46" in 1977. Salinas echoed the pattern, receiving 6.83" in 1976 and 8.02" in 1977 (Salinas has an average rainfall of 13.03 inches).

While annual rainfall is an important factor in providing water, more important is how that precipitation is distributed over the winter months. The drainage basins that provide the water for the city of Santa Cruz overall receive approximately 45 inches of rain on average. Those 45 inches could fall evenly over a six-month period (November through April), equivalent to a rate of about a quarter of an inch every day. Some of this rain would be caught by the vegetation cover and be evaporated, and much of the rest would infiltrate into the soil and percolate into the underlying aquifers. Runoff would be much lower than if that same 45 inches fell in a series of more intense storms.

For the city of Santa Cruz, where surface sources or streams provide about 92% of the water supply, its generally better if the rain falls

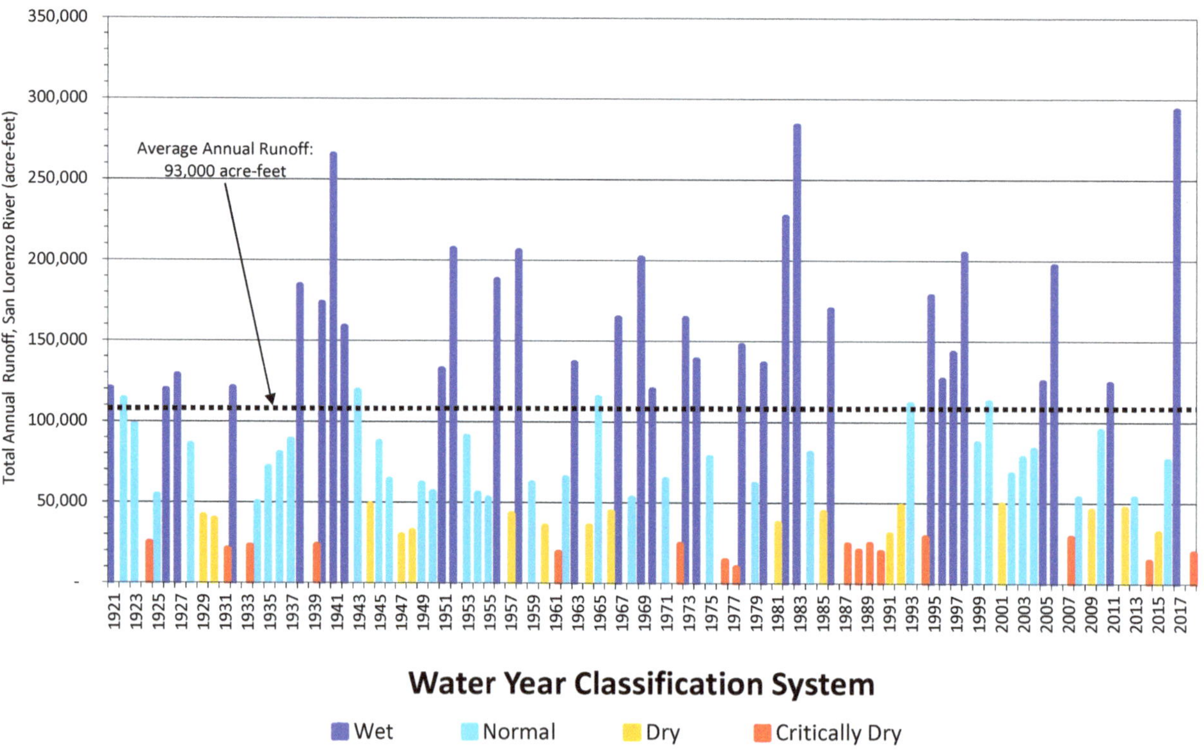

FIGURE 7.7. Total annual runoff from the San Lorenzo River from 1921 to 2017 with designations as wet, normal, dry or critically dry. (Graph courtesy of the Santa Cruz City Water Department).

over a shorter period of time. If the ground surface gets saturated or wet early in the winter, then additional precipitation will run off, find its way into streams and be captured by one of the city's diversions or our single reservoir, Loch Lomond. Stream flow in the San Lorenzo River, our major county stream, is in many ways a better indicator of the severity of drought conditions, as this is the water that is actually available for diversion and providing for the city's needs. Looking at the 78-year history of runoff from the San Lorenzo River, the annual flow from 1976 and 1977 are the two lowest years on record *(Figure 7.7)*. Both were deemed "*critically dry*" with 9,566 and 14,007 acre-feet respectively flowing downstream through the lower river. An acre-foot is an odd measure to try to get your head around, but think of a soccer or football field as being about an acre, and then cover that with a foot of water. One acre-foot is equal to 325,829 gallons of water, or a bit more than two families of four will use in a year. Suffice to say, Santa Cruz was rationing water for both of those years.

In contrast to the Santa Cruz City Water Department, the Soquel Creek County Water District, which serves much of the mid-county, is essentially completely dependent upon groundwater, which is pumped from a number of wells. For this water district, low intensity rainfall spread out over a longer period of time is better for recharging their groundwater aquifer.

THE DROUGHT OF 1987-1992

From 1987 through 1992, California endured one of its longest droughts in recent years *(See Figures 7.5 and 7.7)*. Yearly rainfall over much of the state was about half of the 20th century average. While reservoirs can buffer the lack of rainfall for a year or two, when a drought lasts five or six years, trouble and shortages begin, which lead to restrictions and rationing. These longer dry periods impact everything from agricultural needs to reductions in hydroelectric power production, from reductions of fish populations to declines in groundwater tables as aquifers are pumped more heavily.

By 1990, Santa Cruz County's population had grown to 230,000 while Monterey County had reached 355,000. The drought worsened in 1988 as much of the United States also suffered from a severe lack of precipitation. The six-year drought in California ended in late 1992 as a major El Niño event in the Pacific most likely caused the unusual persistent heavy rains.

Santa Cruz rainfall was below normal for this entire 6-year period and averaged just 19.2"/year, or about two-thirds of average for these six years. The San Lorenzo River runoff for four of those years was ranked as critically dry *(See Figure 7.7)*. Salinas averaged a little over eight inches a year or 63% of normal, while Monterey averaged 14.4", 75% of normal. Prolonged droughts, such as this one, usually lead to progressively lower reservoir levels and also declining groundwater tables, as there isn't enough winter rainfall and runoff to replenish the aquifers and fill the reservoirs. In the city of Santa Cruz, water rationing was again instituted.

THE DROUGHT OF 2007-2009

This three-year period was ranked as the 12th worst drought period in the state's history, and the first for which a statewide proclamation of emergency was issued *(See Figure 7.5)*. This dry period also saw greatly reduced water diversions from the State Water Project, and in some ways was another wake-up call for what was to follow a few years later. Drought impacts were most severe on the west side of the San Joaquin Valley where Central Valley Water Project deliveries were only 10 percent of the farmer's allocations in 2009, following deliveries of 40 percent in 2008 and 50 percent in 2007. These reductions had very significant economic impacts on agriculture and on those rural communities that depend on agriculture for employment. The summer of 2007 also saw some of the worst wildfires in Southern California history.

Overall, precipitation in the Monterey Bay area was 59% of normal in 2007, 84% in 2008 and 84% in 2009. This area didn't suffer during this three-year period to the extent that the Central Valley did.

THE GREAT DROUGHT OF 2012-2016

The 2012-2016 drought is considered by many to be the worst in recent California history, and was one of extreme proportions *(See Figure 7.3)*, with record-high temperatures and record-low levels of snowpack and rainfall, as well as very low river runoff *(See Figure 7.5)*. The five-year duration of the dry conditions, the growing population with its increasing usage and the greater demand for agricultural water all served to make this a very serious event for the state, but one which may become more common in the future as the climate continues to change. Continued warming is expected to lead to: 1) less precipitation falling as snow at higher elevations, which means less snowpack for spring and summer runoff; 2) rainfall more concentrated in the winter months, which translates into more frequent and severe flooding; 3) inability to trap and store as much runoff in reservoirs because of concentrated winter runoff; and 4) warmer and drier summers, which would lead to greater water demands for urban and agricultural users, and also more brush and forest fires, as were experienced in the fall of 2017 and summer of 2018.

By the second year of this drought (2013), California overall received less than 34% of average expected precipitation. For many regions of the state, the rainfall totals for 2013 were the lowest on record. Salinas got just 3.25 inches of rain the entire year (25% of normal), the lowest in its 83 years of record keeping. While the Monterey rainfall record is incomplete and a number of years are missing (since recording began in 1906), the total for 2013 was only 4.13." This is the lowest on record, and 21% of average. Santa Cruz fared considerably better with 18.8 inches, 65% of the average.

As the drought continued, there were a number of direct effects. Stream and river flows were so low that fish couldn't get to their spawning grounds and many river mouths were blocked with sand bars that prevented salmon from even starting their spawning journey upstream. The California Department of Fish and Wildlife estimated that 95% of the winter run salmon didn't survive 2013. An additional impact of the drought and extreme heat was an unprecedented die-off of trees, which increased the risk of wildfires. In November of 2016, the U.S. Forest Service reported that the extended dry conditions led to the death of over 100 million of the state's trees.

In February 2014, the continuing drought conditions led the California Department of Water Resources to develop a plan to reduce water allocations to farmland by 50%. This was a huge issue for the state's 45 billion dollar agricultural industry, which grows nearly half of the nation's fruits, vegetables and nuts. By mid-May of 2014, the entire state was under Severe Drought or a higher-level (Exceptional Drought) condition *(See Figure 7.3)*.

Dendrochronology (or tree ring studies) in the Sierra Nevada by several different groups concluded that the rainfall in California had been abnormally high since about 1600, and that the state had experienced its wettest period in 2,000 years during the 20th century *(See Figure 7.4)*. This climate history does not bode well for California's future and our water dependence.

In June 2015, the governor ordered all cities and towns to reduce water usage by 25%, which is important, but with agriculture in California using about 80% of water, domestic use reduction couldn't solve the water deficiencies. As agricultural water allocations were reduced, pumping of groundwater from wells was increased, which led to continuing decline in Central Valley water tables.

The Sierra snowpack levels were at or near record lows during the winters of 2013, 2014 and 2015, with the statewide snowpack on April 1, 2015, holding only five percent of the average for that date, a record which extends back 65 years. The snowpack normally provides about one-third of California's water supply so this was bad news indeed.

And has often been the case in the past in the state, a very wet winter follows a period of drought, or just the opposite happens, a drought follows a very wet winter. This happened with the floods of 1861, which were followed by the drought of 1862-1864. The winter of 2016–17 turned out to be the wettest on record in Northern California, surpassing the previous record set in 1982–83. Runoff filled Oroville Reservoir (the second largest in California with a 770-foot high dam, the tallest earthen dam in the United States) and overflow down the spillway in early February led to partial failure of the concrete structure, which prompted the temporary evacuation of nearly 200,000 people north of Sacramento. In response to the heavy precipitation, which flooded multiple rivers and filled most of the state's major reservoirs, Governor Brown declared an official end to the drought on April 7, 2017.

The 2016-17 winter rains did, however, fill or partially fill many of the state's reservoirs that had been so depleted during the previous dry years. With yet another reversal in hydrologic conditions, December of 2017 was the second driest in Santa Cruz history. Progressing into 2018, however, both rainfall and snowpack in the state were significantly below average, bringing to mind the drought that had just ended. In August of 2018, Loch Lomond was at 94% capacity, while the much larger San Antonio and Nacimiento reservoirs in Monterey County were at just 17% and 22% of capacity, respectively.

SOME FINAL THOUGHTS ON DROUGHTS

The conditions of recent years, with 16 of the warmest 17 years globally since we began keeping track of temperatures in 1880 all having occurred since 2000, provide strong evidence that the Earth is continuing to warm. This will affect us in California and along the central coast in ways that are not yet completely certain, but the patterns and past history are reminders that the future may well be quite different than the recent past. We will explore this more in the next chapter, which covers climate change.

We can get by without water for about three days if we are just trying to survive, but we all have a desire to do more than merely survive. The paleoclimate history of the southwest, where Bristlecone pines have left us with a record of decades-long droughts in their growth rings, provides some clues as to what the state has experienced in the past. The big difference is that we aren't a small group of Native American people who have learned how to live off the land and can relocate our villages when necessary. Today we are a region (Monterey and Santa Cruz counties) with about 715,000 people, and a state with nearly 40 million people, which supplies about half of all of the fresh produce for the entire country, and it all requires water. California also produces 90% of all the nation's wine, and grapes require water also. While we have built dams, reservoirs, canals, aqueducts and pipelines to move water from where it is to those drier places where it isn't, these vast engineering structures still require occasional precipitation.

We are far more self-sufficient here on the coast, and are not dependent on outside water sources. Nonetheless, we share the same periodic drought conditions that the rest of the state experiences, and populations of both Monterey and Santa Cruz counties continue to increase. This means greater future demands on our water systems and supplies. The big uncertainty is the same as it always has been – how will future climate change affect our precipitation and temperatures? We need to continue to explore all possible options for our future water supplies, although the options are quite limited, and be prepared for the next extended drought. Not just prepared with a report that describes all of the possible options, but with some plan that is ready to implement or a system that can be turned on in a few months time. We are still a long ways from this in the Monterey Bay region, but we can be certain that droughts will continue into the future, and as we will learn in the next chapter, they likely will be more frequent and more severe.

Chapter 8

A Changing Climate

I have a dream, a song to sing... to help me cope with anything
– Benny Andersson and Björn Ulvaeus (ABBA)

INTRODUCTION

Society's need to cope with and adapt to a changing climate and the associated environmental conditions isn't new. Humans have been adjusting to their environment since the dawn of civilization roughly 8,000 years ago. Agriculture is one of the earliest examples: over the ages farmers repeatedly modified cultivation practices and bred new plant and animal varieties that were suited to changing climatic conditions. In recent times, dams and reservoirs, crop insurance programs, and floodplain regulations, are a few examples that reflect our efforts to stabilize and protect our farms, homes, and livelihoods, as well as our food and water supplies in response to a varying climate. The 2012-2016 drought was one of the driest on record in California history, which was followed by the wettest year for the state overall. The Monterey Bay area followed a similar trend and these recent years have provided us with a wake-up call of sorts of what we may expect in the future – unpredictability.

In striking contrast to the natural disasters discussed in earlier chapters, however, climate change is a longer-term process. It is not a sudden or instantaneous event like an earthquake, landslide or flood. Climate is how our particular region or area behaves over the centuries or millennia: our temperature ranges, rainfall and its variations, humidity and other similar phenomenon. Weather, on the other hand, is short term like we hear or read in the news: *what are the high and low temperatures predicted for the day? Is it likely to rain,* etc.

Although the frequency and intensity of weather related events – like floods or severe El Niño winters with heavy rainfall and large waves – can clearly vary over decades as climate changes, we can't honestly look at one year or even several years and proclaim that the storm we just suffered through was driven by climate change. The climate changes over many years and is a long-term process, so one could argue that this discussion doesn't even belong in a book about natural disasters. But we have now measured and observed countless indicators that global climate is changing, and these will surely affect us in the Monterey Bay region over our lifetimes, just as the more immediate or extreme events discussed earlier will. So I chose to step back to discuss the much larger picture by adding a final chapter on climate change and what it may mean for our region.

The continuing warming of the planet has been well documented from thousands of temperature recording stations around the Earth *(Figure 8.1)*. Sixteen of the 17 warmest years on record globally (since 1880) have all occurred since 2000, and 2014, 2015, 2016 and 2017 were the hottest of all. This trend is not just a coincidence. Scientific consensus among climate scientists based on an overwhelming body of evidence indicates that climate change is happening, that it is caused in large part by human activities, and unless urgent action is taken at all levels of government to both mitigate and adapt to it, the human population and our surrounding environment could experience increasingly serious and damaging effects in the decades ahead.

"The greenhouse effect has been detected, and it is changing our climate now."

–James Hansen,
Chief of NASA's Goddard Institute for Space Studies during a U.S. Senate hearing on global warming, 1988.

"We have to deal with greenhouse gases. From Shell's point of view, the debate is over. When 98 percent of scientists agree, who is Shell to say 'Let's debate the science'?"

–John Hofmeister
President of Shell Oil Co., 2006.

"There are three responses to climate change: mitigation, adaptation and suffering. We are already doing some of each. The only question is what the future mix will be. The more mitigation we do, the less adaptation and suffering we will have to do".

–John Holdren
Scientific Advisor to the President of the United States, 2008-2016.

Future climate changes in the Monterey Bay area are very likely to include: a) higher temperatures, more frequent heat waves, and longer droughts with water shortages; b) an increase in brush and forest fires; c) more concentrated rainfall in winter months accompanied by more floods and slope failures; and d) a continuing rise in sea level that will lead to increased coastal erosion and more frequent flooding of low-lying shoreline areas.

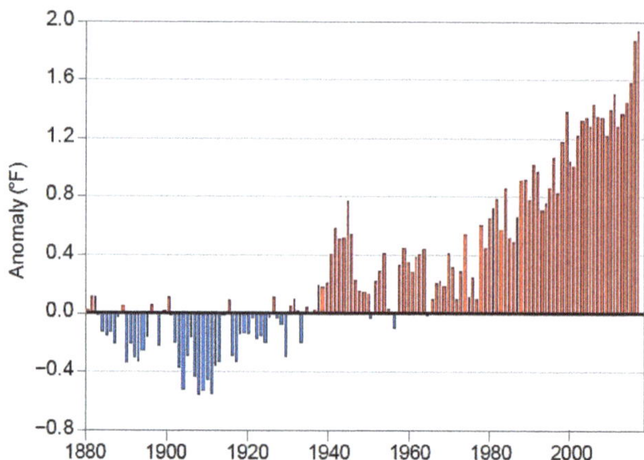

FIGURE 8.1. The average annual global temperature has increased by more than 1.2° F (0.7° C) for the period 1896-2016 relative to 1901-1960. Red bars denote temperatures that were above the 1901-1960 average, and blue bars indicate years with temperatures below the average. (Graph courtesy of the United States Global Change Research Program, Climate Science Special Report).

HIGHER TEMPERATURES, MORE FREQUENT HEAT WAVES, AND DROUGHTS WITH WATER SHORTAGES

A gradually warming climate will have a number of impacts on the Monterey Bay region. Perhaps the change with the greatest influence would be a reduction in the amount of available water. For a region that is already short on fresh water during extended dry periods or droughts (discussed in Chapter 7), these stresses can be expected to increase as populations continue to increase; water demands go up but availability does not. More concentrated winter rainfall and runoff (discussed below) would mean less water available in the later spring and summer months, and for both counties, existing water storage is limited. This is particularly true on the Monterey Peninsula and in Carmel Valley where the San Clemente Dam is now gone and the reservoir behind the Los Padres Dam has lost a significant amount of storage due to sediment fill. Santa Cruz has Loch Lomond, which when full can provide a little over a year of the district's demand if it is the sole source being used. There really isn't a plan B in place at this point for another extended drought in either county, however.

There is an old Chinese proverb that says: "*Dig a well before you are thirsty.*" A healthy person can go about three minutes without air, three days without water and about three weeks without food. Let's assume we can always find good air to breathe and food to eat in our coastal paradise, but water will become the limiting factor over time. Both counties are engaged in discussions and debates about where our future water supplies are going to come from. But we can probably all agree that as long as populations continue to grow around the bay, we are going to need more water than we have access to today, and a changing climate is not likely to improve our situation.

There aren't a whole lot of options out there for additional water supplies at either end of the bay, and for the alternatives being considered, there is no consensus or agreement on the best approach. More conservation, conjunctive or coordinated use of surface and groundwater, reclaiming wastewater and desalting seawater are the most likely options. Importing water from somewhere else is one other option, but given the present demands for water in the Santa Clara and San Joaquin valleys, it is highly unlikely that the Monterey Bay region can ever count on imported water. Each of the above approaches has their limits, their supporters and antagonists, and other than more conservation, they each have significant costs.

Those who oppose any of these potential future water sources will invariably bring up the issue of cost, with the simple statement that "*it is going to cost more than we are paying today.*" And they are absolutely correct, as everything in the future will cost more. What we all need to understand, however, is that the water supply systems in both counties, dams primarily, were built and paid for decades ago. While we still pay for operations, maintenance, upgrades and replacement of pipes and treatment systems, the big-ticket items were built years ago. Any of our options, whether reclaiming wastewater, conjunctive use through piping and then pumping surface water into groundwater aquifers and then pumping it back out again, or desalination, are all going to cost money – a lot of money. There is no more free or cheap water around.

AN INCREASE IN BRUSH AND FOREST FIRES

California has dry, hot and often windy weather conditions in the summer and fall months that frequently produce devastating wildland fires. The state has experienced an increase in the number and size of brush and forest fires in recent years. Fourteen of the 20 largest wildfires in California history have occurred since 2000, and the six biggest have all occurred in the past 15 years. This looks like a clear trend. As of August 2018, the largest fire on record, the Mendocino Complex, had burned nearly 460,000 acres but was finally contained.

The Santa Cruz Mountains and Los Padres National Forest in Monterey County share some of the same characteristics of other coastal mountains in California: dry, windy and hot weather conditions extending from late spring through autumn, which can quickly turn even a modest grass fire into an out of control brush or forest fire. Over the past 33 years the Santa Cruz Mountains have been hit with five large fires *(Figure 8.2)*: the Lexington (1985: ~14,000 acres burned), Croy (2002: 3,127 acres), Summit (2008: 4,270 acres), Lockheed (2009: 7,817 acres), and the Loma fire (2016: 4,474 acres).

The rugged topography in the Los Padres National Forest of Monterey County has a similar recent history with five large fires since 1977, although the acreage burned has been significantly larger due to the steep and almost inaccessible terrain: Marble Cone (1977: 178,000 acres), Indians (2008: 81,378 acres), Basin Complex (2008:

FIGURE 8.2. Recent wildland fires in the Santa Cruz Mountains showing areas burned and year of fire. (Map Courtesy of the *San Jose Mercury News* and Google Earth.)

162,818 acres), Tassajara (2015: 1,086 acres) and the Soberanes Fire (2016: 132,127 acres). At the time, the cost of fighting the Soberanes Fire reached $236 million, the greatest of any fire in the history of the United States.

The steep and often inaccessible terrain in these areas makes fighting these fires difficult, and as more homes are built in more rural mountainous areas, more of the public ventures into this steep rugged terrain, and the weather gets warmer and drier, the probability of large fires occurring more often will increase along with greater losses. Unfortunately, there are really no practical mitigation measures that can significantly reduce the likelihood of these fires continuing to occur.

CHANGES IN RAINFALL DISTRIBUTION, FLOOD FREQUENCY AND SLOPE FAILURES

A warmer ocean will increase evaporation rates, which will in turn increase the amount of moisture in the atmosphere that is available for precipitation. Future rainfall patterns will likely be altered by this available moisture as well as atmospheric pressure differences and storm tracks. Most climate models predict more rainfall concentrated in the winter months, which would tend to lead to more frequent and larger floods as well as more slope failures. Although we have rainfall records extending back to 1868 in Santa Cruz, and discharge records for the major streams going back many decades, these are typically reported or listed as annual values without being broken down by month, so without additional analysis it is not possible yet to determine if there have been any long-term changes in seasonality of rainfall and runoff patterns.

There is a well-documented history of flooding in both Santa Cruz and Monterey counties and the communities and areas susceptible to flooding are not mysteries *(See Chapter 5)*. These will be the same areas affected in the future if flooding intensifies. It is also important to keep in mind that climate is long-term and weather is short-term. It will take a while before we can unequivocally say that any individual event – whether flood, fire or drought – was definitely caused or intensified by climate change, although trends are starting to become increasingly apparent.

The great majority of landslides, debris flows and other mass downslope movements occur during the winter months, with these events usually triggered by prolonged and/or intense precipitation. The most vivid recent example of this was the Montecito (Santa Barbara County) debris flows of January 9, 2018, which followed the largest brush fire in California history. In 15 minutes the steep barren hillsides received about three-fourths of an inch of rain, which generated debris flows that swept rapidly downslope in the middle of the night, destroying over 100 homes, damaging an additional 300, causing at least 21 deaths and injuring 150 more people. Santa Cruz County suffered similar losses during the intense and prolonged rainfall of early January 1983, and the Big Sur village area was devastated by debris flows in the winter of 1972 following the Molera Fire of the previous summer *(See Chapter 6)*. While these are not new events for the Monterey Bay region, there is a high probability that they will occur more frequently in the future.

SEA-LEVEL RISE

Global warming and sea-level rise were nearly unheard of 20 years ago outside of university seminar rooms, scientific meetings and technical journals. Today, however, they are common household words and frequent front page stories. Many climate scientists consider global climate change and sea-level rise as the greatest challenge that human civilization has ever faced. While climate has changed throughout the entire 4.5 billion year history of the Earth – it's been much warmer and much cooler than today – through all but the last 8,000 or so years there was no human development that was affected by those climate changes.

When the last ice age ended about 18,000 years ago, much of the far northern hemisphere was still covered with ice (Puget Sound and the Seattle area, for example, were covered with about 3,000 feet of ice!). All of the water locked up in that ice came from the evaporation of ocean water, which lowered sea level around the world about 400 feet. As a result, the shoreline of Monterey Bay was eight to ten miles to the west at that time, out at the edge of today's continental shelf *(Figure 8.3)*. As the ice age ended and the climate gradually warmed, two processes conspired to raise sea levels around the world: the water in the oceans warmed and expanded, and ice sheets and glaciers melted with that meltwater flowing into the oceans.

From approximately 18,000 to 8,000 years ago, sea level rose relatively quickly at average rates of about a half an inch a year or five inches every decade *(Figure 8.4)*. This gradually moved the Monterey Bay shoreline landward or inland *(See Figure 8.3)* at four to five feet every year on average. Coastal habitats, whether kelp beds, marshes or the intertidal zone, as well as sand dunes and beaches gradually migrated landward, keeping up with the gradual rise in sea level. While there may

FIGURE 8.3. A 3-dimensional bathymetric map of Monterey Bay shows the continental shelf in light blue that was exposed as dry land during the last ice age 20,000 years ago. The shoreline at that time was where the shelf drops off into deep water today. (Illustration courtesy of the Monterey Bay Aquarium Research Institute).

well have been Native Americans living in this area during the latter part of this period, they had no permanent settlements so a migrating shoreline was not a concern for them.

Sea-level rise leveled off about 8,000 years ago *(See Figure 8.4)* and then rose very slowly until perhaps 150 years ago, when climate began to change as a result of human activity. The

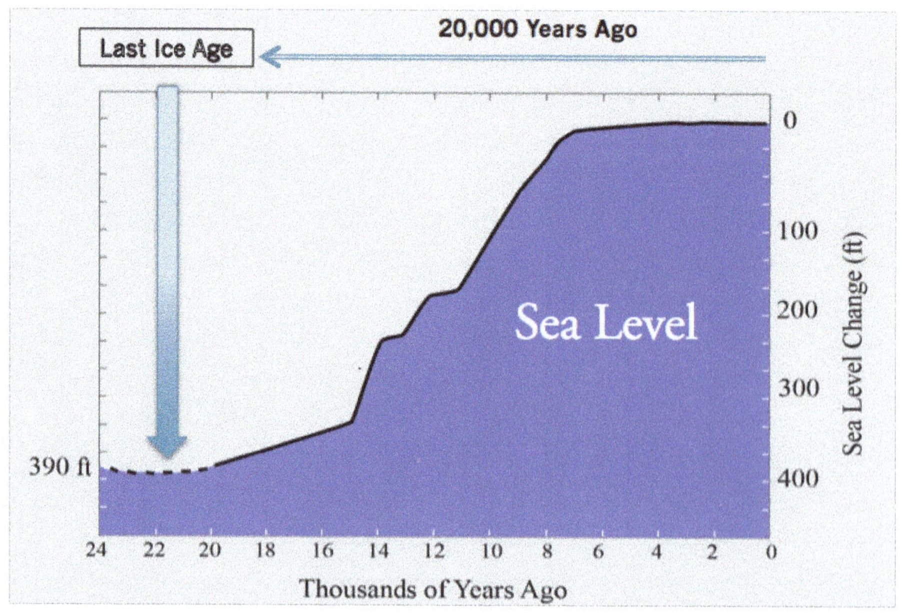

FIGURE 8.4. Sea level was about 400 feet lower 20,000 years ago at the end of the last ice age and then rose relatively quickly until about 8,000 years ago when it leveled off and rose very slowly until about 1900. (Graph courtesy of John Englander, *High Tide on Main Street*).

FIGURE 8.5. During the strong El Niño of 1983, elevated sea levels combined with high tides and storm waves overtopped the seawall along Beach Drive in Rio Del Mar.

onset of the Industrial Revolution and the increased burning of fossil fuels (coal initially and then oil and natural gas) led to a gradual rise in the amount of carbon dioxide in the atmosphere, which has continued to increase over time. Carbon dioxide (as well as methane, nitrous oxide, ozone and halocarbons) was recognized as a greenhouse gas well over a century ago, and by trapping the heat leaving the Earth – much like rolling up your car windows on a hot day – the planet has been gradually warming. Glaciers are retreating and the huge ice sheets of Antarctica and Greenland are melting, which continue to add more water to the oceans.

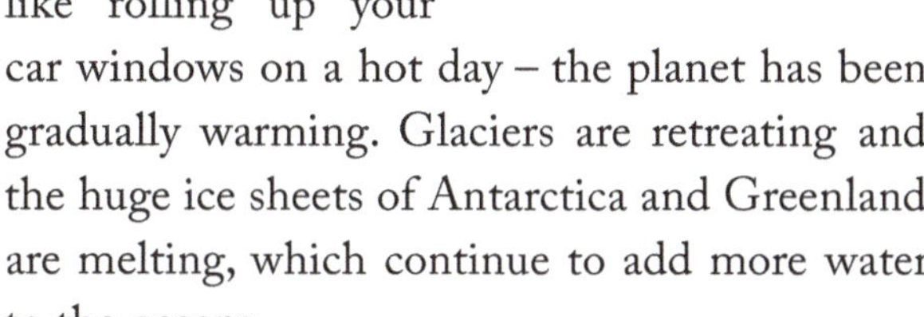

Sea-level rise is probably the effect of climate change that has generated some of the most visible impacts along the Monterey Bay coastline historically and that will continue to produce some of the most significant effects along the shoreline in the future. A continuing rise in sea level, and very likely at an increasing rate, will lead to several types of impacts, which we have already begun to experience. While low-lying coastal areas already experience flooding during periods of very high tides, large storms and El Niño events *(Figure 8.5 and Chapter 4)*, this will occur more frequently, followed in time by permanent inundation of the lowest-lying shoreline areas, just as occurred between 18,000 and 8,000 years ago.

Our longest records of recent sea-level rise along the California coast come from official NOAA (National Oceanographic and Atmospheric Administration) tide gauges or water level recorders that have been installed at 12 different locations along the state's coast. There are two of these gauges that provide a record of what the central coast has experienced, one in Monterey and one a bit farther away in San Francisco. The record for Monterey is relatively short, extending back only 45 years, but nevertheless has documented an average rise in sea level of about 1.48 mm/year since 1973, equivalent to 5.8 inches/century *(Figure 8.6)*. San Francisco, on the other hand, is the longest running water level recorder in the nation, having been installed in 1855. It reveals a slightly higher average rate of sea-level rise, 1.96 mm/year or 7.7 inches/century *(Figure 8.7)*. For the 90 miles of coastline between these two stations, including Moss Landing and Santa Cruz, we don't know exactly what has happened,

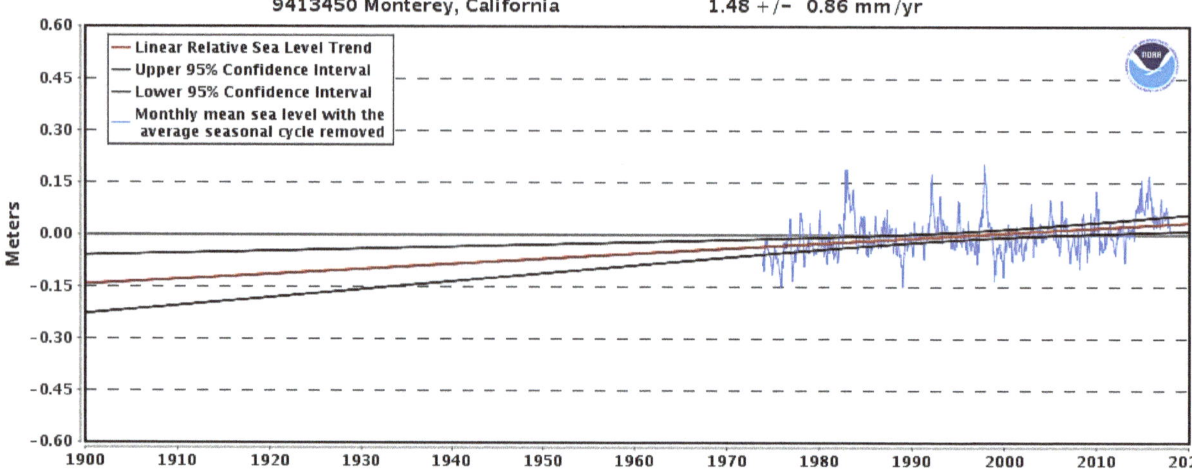

FIGURE 8.6. The official NOAA tide gauge in Monterey has been monitoring sea levels since 1973 and has recorded an average sea-level rise rate of 1.48 mm/year, equivalent to 5.8 inches per century.

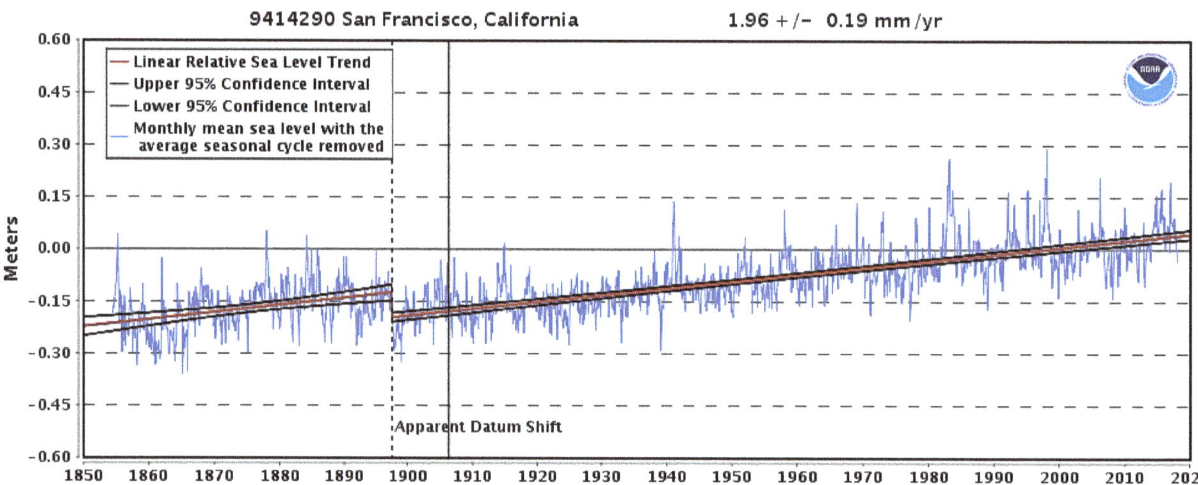

FIGURE 8.7. The NOAA tide gauge at the Golden Gate in San Francisco is the oldest in the United States and has recorded an average sea-level rise rate of 1.96 mm/year or 7.7 inches per century since 1897.

except that the rate of rise is probably not too much different and likely lies between these two values.

You may be wondering why sea level isn't rising at the same rate in Monterey and San Francisco, which is a logical question to ask at this point. While the overall volume of the oceans determines the global elevation of the sea surface or sea level, what is measured at any particular location with a tide gauge actually reflects two things, both global sea level but also the motion of the land where the gauge is located. Measured sea-level rise can, therefore, vary from place to place. The tide gauge at Monterey is mounted on Fisherman's Wharf while the San Francisco gauge is located on the National Park Service wharf, which are both attached to the land.

Virtually every coastline around the world is geologically active to one degree or another, and may be slowly rising or sinking, or in some places may be relatively stable. As a result, every tide gauge will be measuring the combined effects of how the ocean surface is changing over time and also how the land is moving. Looking at the sea-level rise records around the coastlines of the United States shows some large differences from Alaska to New Orleans. Alaska was covered with thousands of feet of ice during the last ice age, which put a huge load on the land, causing it to subside, much like sitting on your mattress. When the glaciers melted and retreated the land began to rebound, although a whole lot slower than your mattress does when you stand up. It's taken thousands of years and it is still rebounding. As a result the coastline for much of Alaska is actually rising faster than sea level, such that relative to the land, sea level is actually falling. The tide gauge at Skagway, Alaska, for example, shows a drop in the level of the ocean of 17.6 mm/yr. or 5.78 feet/century. To state the obvious, the people living in Skagway aren't concerned much about sea-level rise.

At the opposite extreme is the area around New Orleans, which is subsiding as a result of the load of thousands of feet of Mississippi River sediments on the Earth's crust, but also from the extraction of petroleum and groundwater and the resulting sediment compaction. The tide gauge at Grand Isle near New Orleans records a much higher local rate of sea-level rise, 9.6 mm/year or the equivalent of 3.17 feet/century, three times the global average. If the residents of New Orleans didn't have enough to worry about with hurricanes and flooding, they need to think seriously about sea-level rise as well, and how they are going to respond in the decades ahead.

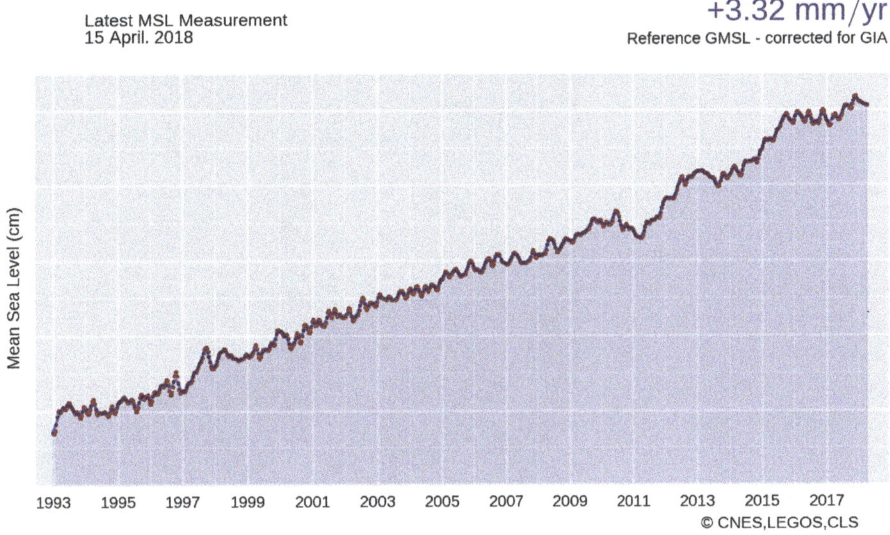

FIGURE 8.8. Satellites have been recording global sea level from space since 1993, which has averaged 3.32 mm per year or 13 inches per century and this rate is increasing. (Graph courtesy of the AVISO CNES Data Center).

What is clear is that the rate of global sea-level rise is increasing and it has been precisely measured from satellites since 1993, which produces absolute or global values, free of the influence of local land movements *(Figure 8.8)*. The global rate has more than doubled from about 1.4 mm/yr. (5.5"/century) during most of the last century to an average of about 3.32 mm/yr. (13"/century) over the past 25 years. All indications from the continuing increase in greenhouse gas emissions is that this rate will increase for many decades into the future. The question that is very difficult to answer, is now much the rate will increase? This is closely tied with how much more carbon dioxide goes into the atmosphere in the decades ahead, which is directly related to how much more oil, coal and natural gas we burn.

Those low-lying areas around the shoreline of Monterey Bay that are flooded now during El Niño events, King tides, and the simultaneous occurrence of very high tides and storm waves (discussed in Chapter 4) are the same areas that

FIGURE 8.9. The El Niño of 1983 combined with repeated large storm waves arriving at times of high tides through the first three months of 1983 washed over and closed low-lying portions of East Cliff Drive carrying large logs with it.

FIGURE 8.10. Sand and debris were washed over the seawall and into downtown Capitola by large waves arriving at times of very high tides during the 1978 El Niño.

will be flooded in the future with higher sea levels, although the extent of inland intrusion of ocean water and the frequency of these events will continue to increase. Some of the low-lying areas around northern Monterey Bay that are particularly susceptible and will see more future flooding include portions of East Cliff Drive at Twin Lakes, Corcoran Lagoon and Moran Lake *(See Figures 4.45 and 8.9)*; the oceanfront areas of downtown Capitola and Rio Del Mar, as well as Seacliff State Beach and Beach Drive *(See Figures 4.44, 8.10 and 8.11)*.

Higher sea levels will also lead to waves breaking up against eroding sea cliffs more often, which will increase the rates of coastal cliff retreat. West Cliff Drive, East Cliff, Opal Cliffs,

FIGURE 8.11. Waves at high tide overtopped the seawall and flooded the Esplanade in Rio Del Mar during the 1983 El Niño winter.

and Depot Hill are areas that experience the greatest exposure to wave erosion now and this can be expected to gradually intensify with higher future sea levels. The sandy shoreline of southern Monterey Bay has been suffering from bluff retreat for decades *(See Chapter 4)* and we can expect that this will continue into the foreseeable future, and likely at increased rates.

SOME FINAL THOUGHTS ON THE IMPACTS OF CLIMATE CHANGE

The climate is changing and there is strong scientific agreement that human activities – primarily the burning of fossil fuels and also the production of cement – are the most important causes. The processes that will be affected or enhanced by climate change are not new to the Monterey Bay region, but they are very likely to become more frequent and be of greater magnitude, whether floods and landslides, droughts and forests fires, or sea-level rise. These are all driven by global climate change, something that we as individuals cannot exert a great influence on. It will require mitigation through a major worldwide reduction in our emission of greenhouse gases. However, we have a very significant role in California as the world's 5th largest economy and a leader in responding to climate change. We have influence well beyond our state's borders as many other states and nations often take their cue from us regarding auto emission standards, targets for transitioning to renewable energy sources and aggressive pursuit of solar and wind power.

In addition to our efforts in climate change mitigation, we also need to focus our energies and planning on adaptation, simply because there is a large amount of climate change and warming already built into our atmosphere from past emissions. Despite our best efforts and continued efforts to reduce our use of and dependence on fossil fuels, we are going to experience the impacts of climate change for decades into the future. We can reduce these but we need to plan for future water shortages, sea-level rise and the other impacts discussed above. We need to dig a well before we get thirsty and begin figuring out how we can respond and prepare for the inevitable changes that are coming.

Index

www.ingramcontent.com/pod-product-compliance
Lightning Source LLC
Chambersburg PA
CBHW042355030426
42336CB00030B/3493
9781732709300